# *Running a Tack Shop as a Business*

# *Running a Tack Shop as a Business*

## JANET W. MACDONALD

J.A. Allen
London

British Library Cataloguing in Publication Data
A catalogue record for this book is available from the British Library

First edition published 1986
Revised edition 2000

Published in Great Britain by
J.A. Allen
45-47 Clerkenwell Green
London EC1R 0HT

ISBN 0 85131 824 X

Design by Judy Linard
Typeset by Textype Typesetters, Cambridge
Printed by Dah Hua Printing Press Co Ltd., Hong Kong

# Contents

# Acknowledgements

My grateful thanks to all the people who helped me by providing information for both editions of this book: Anthony Wakeham at the British Equestrian Trade Association, Barry Fehler at South Essex Insurance Brokers, my accountant Philip Murphy and all the manufacturers, wholesalers and tack shop proprietors who tolerantly answered my questions, but especially the late Norman Simmons, Freddie Hughes and David and Pam Dyer. And as always, to my husband Ken Maxwell-Jones for information on personal finance, his tolerance of my temper when writing, and unending supplies of coffee!

## Author's Notes

I don't know anyone called Bloggs who runs a tack shop or saddlery. If there is one, I apologise for using their name and hasten to add that I intend no mischief, for it was just plucked out of the air for illustration purposes.

In these days of political correctness, an author has to contend with the problem of how to use pronouns. One either risks offending some readers, or has to use the cumbersome alternatives of plural pronouns or the even more cumbersome form of 'he or she'. I dislike both, and have thus used the male and female pronouns at random, with no intention of suggesting any correlation between occupation and gender.

# Introduction

Some years ago, when I was writing the first edition of this book, I observed a fascinating little saga. Close to my home, in the middle of an affluent and very horsy area, and at least ten miles from any saddlery, there was briefly a tack shop. It lay on a main road, en route to the livery stables where my horse lived, so I passed it every day and I saw it open then rapidly close. It started when paint was applied and a sign saying 'Open for feed NOW, main shop opening soon' appeared in the window. Then bridles could be seen hanging above the curtain in the window, but before I had a chance to go in and buy something it was closed - all in a couple of months.

The sign saying 'Lease for Sale' remained for over a year. I was never able to find out exactly what happened, although I did hear a rumour that the proprietor had complained that the customers in this area were too fussy. That may or may not be true, but what is obvious to me is that he made a serious mistake somewhere. An expensive one too, for he was lumbered for many months with a lease and insurance on premises which were earning him nothing.

There is a simple rule in business, and it hasn't changed since mankind started taking goods to market - get it right and you'll make a bit of money; get it wrong and you'll lose a lot.

The sad thing is that there are too many people in the saddlery business who get it wrong, in a variety of ways, and end up out of business and in financial difficulties.

The worst of the classic mistakes is the assumption that a knowledge of horses and saddlery is more important than a knowledge of modern business methods – or knowledge of business at all. Business is essentially about money, and if you are not prepared to keep a close eye on the money side of your business, you should give up the idea and get a job where someone else does the worrying and pays you a wage to do what you are good at.

Another mistake is the failure to understand that the retail saddlery trade is very capital intensive. This means that you have to carry a lot of very expensive stock, which ties up a lot of your capital. If you cannot earn considerably more from this stock than you could from investing your capital elsewhere, there isn't a lot of point in doing it at all.

These are only a couple of the many points that need consideration before you set up shop. The purpose of this book is to help both established shop-owners, as well as those thinking of starting a shop, with these considerations, and to show them how to avoid all the pitfalls that lie on the road between failure and success.

Success always means hard work, and running your own business means long hours as well. You don't stop work when you shut up shop at night, for there is book-keeping to be done, stock to be ordered, bills to be paid and plans to be made. It won't be easy, but the satisfaction of working for yourself and the opportunity of gaining financial independence are always worth the effort involved.

(There are some special considerations for those thinking of adding a shop to an existing riding complex and I have dealt with them in a separate chapter.)

Since I wrote the first edition of this book there have been many changes in the tack shop business – indeed, I could almost say that the business has changed out of all recognition. There is now a big fashion element to both riding clothes and equine equipment, which means the retailer has to be even more aware of stock levels, and many new manufacturers have entered the UK, from as far afield as America and the Antipodes as well as our European neighbours. On the negative side, there has been a marked increase in all aspects of theft, and rural businesses are now just as vulnerable as those in towns.

*Introduction*

With this in mind, I would urge all retailers to join the British Equestrian Trade Association; there can be no better way to keep abreast of the trade. They will only accept you as a member if you are a bona fide retailer operating from a proper shop; but once you are in you will be part of an organisation which is prepared to help, give advice and fight for your interests. You will be able to attend BETA courses on various aspects of the business, including one that will give you the qualification you need to sell wormers. They publish a quarterly members' newsletter, run publicity campaigns and a gift voucher scheme, and issue an address list every year of all their members, both retail and wholesale. They also hold an excellent trade fair every year where you can go and see what new products are available and meet the manufacturers and wholesalers, and they also publish a booklet called *What to Wear* which will help you answer queries on correct turnout for show classes.

# 1

# *Finding and Financing the Right Place*

## Location

Once you have made up your mind that you want to run a tack shop, you must find suitable premises. The ideal place will be in an area where there is a high population of horses and where there is little or no competition. It is going to be difficult enough getting your shop going without all the aggravation that can come from an unfriendly competitor. It is not unknown in such circumstances for an established shop to sell its goods at a loss for long enough to squeeze out the newcomer. Even if they don't do this, there may not be enough trade to support more than one shop, so all that happens is that neither of you makes a living. To avoid this problem, you should be at least ten miles apart.

It is always easier to set up business in an area where you know the local conditions and where you are known yourself. The latter is especially valuable where the bank is concerned – but more of that later. Often, it will be the local grapevine that tells you an existing tack shop is for sale before it is advertised; and as often as not it is the availability of such a place that propels the ambition to have your own shop from the back to the forefront of your mind.

You will need to go and have a good look at it and perhaps ask some questions of the locals which will help you in your decision. Be particularly wary of any business that has a poor reputation locally; although you will obviously notify suppliers of the change of ownership, and place adverts in the local papers to let other people know, it can take years before such a legacy wears off.

## Background to the Business

There are a lot of questions that you must ask about the background to any business.

The most common motive for hanging out the 'For Sale' sign is that business is bad, so the first and most important question is: 'Why is this place available?' The next should be: 'How many people have had it in the last ten years?' If the answer to the second question is more than two, you have the answer to the first at hand. For some reason, it is not profitable, and you would be wise to forget it. Are you really so special that you can make a go of a business where others have failed? There are, of course, other reasons why a place is available that may make you decide against it, such as an imminent motorway between it and all the local stables. This is something that other local businesses may know about and it is worth enquiring from them yourself before you pay a solicitor to check at the Land Registry.

It is rare to find a tack shop offered as a going concern, but if this is the case, there is a separate set of questions to be asked. Do the seller's claims about the business tally with his actual accounts/tax returns? Is the price he has set fair and reasonable? For instance, what value has he placed on the goodwill; are there any outstanding bills for you to pay, or others for you to receive – and how likely are you to be paid? Be sure that his rates and taxes are up to date, and that there are no staff redundancy costs outstanding, especially if it is a limited company you are buying!

What is the position with the existing stock and shop fittings? Normally one deals with these as a separate item from the other aspects of the purchase (lease/rent/goodwill etc.). Independent stock auditors will come in and produce a valuation based on wholesale prices and if you want all the stock you accept that valuation. If you don't want any of the stock you don't have to have it. If you want some but not all of it, you haggle.

Why is the proprietor anxious to sell? Here is the 'motorway' type of problem again, or maybe the bank is about to foreclose on his mortgage. Often the explanation given for the sale is that the proprietor wants to retire, or has some family-related reason to sell. Don't let the plausibility of such reasons prevent you from making all the usual checks. And finally, does he actually own what he is offering you? This last point should always be checked by a solicitor.

What is the parking situation? Remember that many of your customers will be buying bulky items such as rugs or saddles and will not relish a long walk back to their cars. Many of them will have large vehicles like Range Rovers and some may even want to bring a horse in a trailer to try things on him on the spot, or show your resident repairer what is needed in the way of saddle stuffing.

## Finding Premises

If you are not looking at the prospect of taking over an existing tack shop, you will have to start from scratch. There are many ways of finding suitable premises, from perusing the small-ads to using a shop broker. As well as looking in the classified section of *Equestrian Trade News*, try your local paper, *Dalton's Weekly*, or *Exchange and Mart*. Drive around your chosen area and look for suitable premises with a 'For Sale' sign, or even unsuitable premises to tell you which estate agents deal in such properties.

Business transfer agents (find them in the Yellow Pages) will have a list of likely properties and they will also provide stock auditors, offer to help you raise money and help in other ways if you are inclined to let them. Just bear in mind that they gain their main fees from the vendors, as do estate agents, and make sure that there is no conflict of interest between that and the service they offer you.

Whichever way you find suitable premises, you must be

prepared to pay a lump sum down in addition to the lease/rent. This is called either 'key money' if the shop is vacant, or a 'premium' if it is still trading. It is based on a multiple of the weekly takings figure. Anything between thirty and forty-five weeks is normal.

Before you go any further, you must check the situation on planning permission. Just because the shop is currently selling tack doesn't mean that it has permission to do so, and you may be forced to close down or revert to the permitted use, which could be something in which you have no interest. If it is not currently a tack shop you must find out if you can get permission for change of use.

It is also wise to write a courtesy letter to the local Chamber of Trade/Commerce. They take a keen interest in the types of trade in their area and also in what they call 'undue competition', and they may lodge objections to your application for planning permission.

Once you are happy about these aspects, you can ask the last question, which is perhaps the most crucial: 'How much is it going to cost me to run?' Failing to check this has caused the bankruptcy of many inexperienced business people. If you are borrowing money, you must take the interest into account. If the premises are in need of redecoration or major shop-fitting alterations, you must take the cost of these into account. And finally you must check on the overheads, rates, insurances etc. Rates are a particularly crucial point. They may be very high and that will do nasty things to your profit margin if you have to compete with someone over the borough border whose rates are less.

## Raising Finance

All these questions centre on the important points, 'Can I afford this place?' and 'Can I make a profit from it?' If the overall answer is 'No', or only 'Maybe', perhaps you should look for somewhere else.

If, after all this, you decide you do want the place, it is time to think of money. For straightforward renting this is a fairly minor point, although you should keep in mind that your income from the shop may be low for a while and you should have enough money available to pay the rent for at least six months. Do get a solicitor to check the terms of tenancy, to be sure they allow you to do all the things you want to do. You may not be allowed to sub-let for instance, and might be thinking of arranging a repair service on this basis. You should also check the minimum time you must pay the rent. It may be for a fixed term of a year or more, regardless of whether or not you are using the premises and you may have to pay several months in advance at the beginning.

When you are starting up, the rental method is best, but if you cannot find rented premises, look for a short-term lease or one that is renewable at three- or five-year intervals. If you don't make a go of it, you won't be stuck with a long lease. If you do succeed, you may want to move into bigger premises and again find the long lease a burden. On the other hand, if you intend to use the lease as security for a loan, a short lease will not be popular with the lender.

If the lease must be purchased and you do not have the cash handy, or a kind, rich uncle you can tap for a loan, you will have to go to a commercial institution to apply for a loan. Prices vary tremendously according to location, facilities and many other factors. Add to the cost of the property the amount you will need for stock and running costs for the first year or so and you will have to find a substantial amount of money. One problem is that no bank or other source of money will lend you the whole amount. You will be expected to put up at least a third yourself, and possibly half. Banks call this your 'venture capital', and while it is comparatively easy to find it through venture capital brokers in the USA (where the tax laws are kinder on this sort of thing), in the UK it is very difficult.

Building societies are generally reluctant to provide money for business loans, so you must go to a bank or finance house and they will need quite a lot of convincing before they let you have the money. Their first concern will be with you personally and with your past financial history. This is where going to a bank that knows you helps. If this history is steady and sensible – no unauthorised overdrafts, no living hand-to-mouth or other signs of inability to handle money – they will then consider you as a business person. Have you ever run this sort of business or any other before, and if so, how successfully? Have you any qualifications? This won't help much if you have no practical experience to back them up. How much business acumen do you have? This does not mean you have to be able to read a balance sheet, but you should have prepared a simple cash flow forecast (see Chapter 4) showing expected income and expenses for two or three years.

Next they will consider the business itself. Is it to be a new venture, or are you taking over an existing one? Has it been consistently profitable? They will want proof of this in the form of the audited accounts for several years. What are the premises like? Have you had them valued? Have you plans and estimates for any improvements? Are you taking over the existing stock? If not, do you also need a stocking loan? They will want to see a priced list of the items you need as initial stock.

What all this boils down to is that they want to be satisfied that you can make a success of the venture, and repay them regularly without causing them any trouble. Once they are convinced of this, they will want some security on the loan, either a charge on the property or on your house. They will also want you to insure the property and stock in a form acceptable to them.

Finally, the period of the loan, interest rate and repayment periods will be set. The loan will almost certainly be for a set period, usually between five and ten years. Repayments will probably be at monthly intervals.

Not all finance houses are prepared to make 'small business' loans, so you may have to look round for the right one. If you do discover one, you will find that their interest rates are likely to be at least 1.5% higher than a bank's, so on the whole you are better off going to a bank anyway.

Start with your own bank. If they won't lend you the money, ask them why not. If the answer is the lame excuse

18

that you want a sum over the manager's limit, or that he has lent his quota for the year, or (if he is honest enough to tell you) that he doesn't fancy the proposition, you are then at liberty to go to a bank that does. In this case you will be expected to move all your banking business across.

At this point, rather than start yourself on the long traipse round all the local banks and finance houses, you would do better to enlist the help of a financial expert who does it all the time. By this I mean one of those much maligned people – a life assurance salesman. Don't shy away from the idea – wherever you borrow money, you will have to insure your life as part of the loan agreement. The life salesman's reward for doing the leg-work in finding you the money you need is the commission he earns on the life policy. And you can and should leave it to him. You'll only queer the pitch if you try as well. His local contacts will know that he won't offer them a poor proposition, and the fact that he presents it to them is a point in its favour. He will start with his local contacts and move farther afield if necessary. He may well know of a bank manager in some other part of the country who is amenable to such cases.

Some insurance brokers specialise in this type of business, as do many 'tied' agents. These latter work for the big unit-linked life assurance companies. Find the companies from their advertisements in the Sunday papers, find their local office in the phone book and ask for an appointment with one of their senior agents who specialises in small business loans.

I mentioned interest rates earlier. They work like this. There exists a thing called base rate, or bank base rate, sometimes called minimum lending rate. This goes up and down according to the state of the economy, or the government's budget requirement. My bank manager said that he would expect interest of 4% over base for this type of loan, so if the bank rate is 5%, you will be paying 9%; if the bank rate is 10%, you will be paying 14%. This does not mean that you will be paying the same rate of interest for the whole of the period of the loan, set when you take it out, but that as the bank rate fluctuates, so does your rate of interest.

However, you will almost certainly be paying off some of the principal (original sum loaned) as you go, so the amount of interest will become less as time goes by. Interest is calculated daily on the amount of the principal outstanding, so if you should find yourself with some money to spare, it might be worth considering making an extra payment against the principal.

Once the loan is approved, the finance house gives you a cheque, or the bank opens a loan acount for you and transfers the money to your current account, and you are in business. In either case, they will be happiest if you sign a standing order for the repayments and woe betide you if that money is not available on time. Two weeks late and they will start writing rude letters – a few more weeks and they will get your deeds out of the vaults and start looking at them with an acquisitive gleam in their eyes!

# 2
# *What to Sell*

Before you start ordering stock you will need to have a long, hard think about what you are going to sell. Perhaps that should read 'what will sell', for there is no point in stocking your shop with items that nobody wants. The average tack shop turns its stock over two-and-a-half times a year, which means that if your total stock has a retail value of £100,000, your annual takings will be £250,000. For every item that sits on your shelves all year without moving, you need another that sells five times a year, and the more unsaleable items you have, the harder the rest of the stock has to work. And the more each unsold item is costing you, day by day, as the earning capacity of its wholesale price wastes away.

There will inevitably be items which, for various reasons, will not sell easily. To a certain extent, only experience will tell you what they are, but you can minimise your idle stock by considering who your customers will be and deciding where the emphasis of your business should lie.

## Clothing and Saddlery

Firstly, do you intend to concentrate on the horse or the rider? Clothing does turn over faster, but unless there is a smart local riding school to provide a big market for good riding clothes, you will not find it easy to survive on clothing alone. Equally, horse clothing and saddlery, although tending to be the more costly items, turn over more slowly. Nor does it necessarily follow that a big price means a big profit. What you really need is a good mix of both horse and rider items to keep the money coming in.

But if you decide on horse emphasis, you should stock a good variety of costly horse items like saddles, bridles and rugs as well all the day-to-day items like brushes and saddle soap, and a restricted range of rider clothing – rubber boots, black hunting caps, jodhpurs, anoraks and body-warmers. If you want rider emphasis, you keep the horse items basic – just a couple of general purpose saddles, a few basic bridles, plus day-to-day items – and you add coloured caps and coloured breeches, fashionable shirts, hacking jackets and show jackets; bowlers and top-hats, leather boots and lots of gloves, together with books and giftware (of which more later).

Part of the thinking behind this emphasis will be your own personal inclination, but most of it should be based on your research into the horsy population of the area (and the emphasis of your nearest competitor). Check where the competitive orientation lies, because showing and dressage people will scorn you if you fill your shop with the complex nosebands, martingales and bits that show-jumping people like, while the jumping people will go elsewhere if you don't stock them. If you are a dressage purist yourself, banish the word 'gadgets' from your vocabulary and find out what these things are for, so that you can talk knowledgeably about them. Remember, you are not doing this to change the world, you are in it for the profit, which means supplying what the customer wants.

The second consideration is the social and financial standing of your prospective customers. Down-market usually means show-jumping, perhaps a bit of driving, cheaper goods and cash. Up-market usually means eventing, dressage, hunting, showing, expensive and good quality goods, but it also means insistence on a wide range of goods to choose from, including fads and fashions which could leave you with unsaleable stock when the fashion has moved on.

## *Peripheral Items*

Next, are you situated where there is a steady passing flow of non-horsy people? What can you stock that they might buy? Anoraks and wellington boots are universally useful and, oddly enough, black wellies are not that easy to find. Even the most unhorsy family may contain a yearning little girl and she will want a grooming kit, books, posters, model horses or a 'Thelwell' mug.

You will also find that many of your horsy customers will be accompanied by uninterested parents or boyfriends and it is worth considering what you can stock to interest them while they wait. And of course, horsy people themselves do actually have other interests (honest), so it is worth catering to them. One of the best ways of doing this is with books. I don't suggest you should set up a whole book department, but a couple of shelves of basic books on cookery, gardening, wildlife, plant recognition and various country pursuits as well as some horse books, should move steadily.

A high proportion of horsy people also keep dogs, so you could add leads and collars, grooming equipment, toys and beds (and books). An even higher proportion of the riders in this country are female, so why not stock handbags, headscarves and some pretty sweaters?

If you are located away from a big city, you could add all sorts of items connected to country pursuits and sports. Walking sticks, pocket knives, binoculars and shooting sticks are all likely items; even guns and shooting equipment, although you will have to have someone with considerable expertise to look after this department and there are a lot of additional legal constraints as well.

There is a whole series of items that come under the heading 'giftware', from coffee mugs with the Thelwell characters and 'official' posters of famous show-jumpers, to pottery, porcelain or brass ornaments. Until you have a chance to assess the demand for such items, you would be wise to keep stocks to a minimum other than just before Christmas.

# Veterinary Products, Hardware Items and Feed

Back now to some more specifically horsy items. You will find there is a steady demand for 'veterinary' products, so whatever your emphasis you should stock them. Liniment, wound powder, ointments and other dressings, poultices, tendon packs, bandages and sticking plaster are all easy to obtain and no problem to sell. Wormers, however, are another matter. While they are in constant demand, if you are to sell them, you must attend a course to obtain the correct qualification, and comply with a code of practice.

There are a few items of hardware you may want to stock if you have the space, as they are not readily available to customers elsewhere. These include hay racks, feed bins, bucket holders and saddle racks. If you do have plenty of space you could add wheelbarrows and corn bins, and it is well worth making space for brooms, forks and buckets as long as your prices compare well with local hardware stores.

Food supplements or minerals/vitamins are no problem as they come pre-packed and in fairly small quantities. Feed itself is heavy, bulky, deteriorates quickly (this is called 'having a short shelf-life') and attracts the attention of rats – and officials from the Environmental Health Department and the Weights and Measures Department. To keep it, you may need additional planning permission and a specially constructed store. All of which would be tolerable but for the fact that the profit margin on feed is much less than the norm on other equine stock. The usual mark-up is about 20%, and you will need to turn your stock over ten or twelve times a year instead of the usual two-and-a-half times. It is not worth selling feed on anything other than a large scale, and to do so means that you will probably have to provide a delivery service.

## Shop Fittings

Whatever you sell, it will have to be displayed, so you will need some shop fittings. If you sell any form of clothing, you will want somewhere for customers to try it on and you will need a biggish mirror situated so they can get a long view of themselves. You will also need some clothing rails, some shelves and a chair for people to sit on while they try on boots.

For the horse items, you will need some shelves, racking systems, counters to hold open boxes of small items, hooks or rings on which to hang bridles, bits and leather items, and some saddle stands. One of these must be sturdy enough to allow the customers to sit on it with their feet off the ground. The others need not be so sturdy - one good idea I have seen in a big shop is to use a jump as a display stand.

Which brings me to an important rule - never refuse a sale. If someone wants to buy a 'display' stand, sell it and get yourself another. Don't dress your window so you can't get things out of it if someone wants them. If anyone wants something you haven't got, order it for them but be sure to take a deposit so you won't be stuck with it if they don't come back. Finally, take the trouble to make seasonal displays and to move stuff around so that regular customers have to look for it - they might see something else they fancy in the process!'

It is not wise to spend a lot of money on shop fittings when you start out, for in the case of failure, their resale value is minimal. You can buy all manner of beautiful fittings and display stands from specialists, but you can also make a lot yourself. You can also buy shop fittings cheaply second-hand from your local auction room or from the specialist dealers who advertise in the *Exchange and Mart*.

## Obtaining Stock

So where do you get your main stock from? There are two main sources - manufacturers or wholesalers. They all have the same trade terms. They will want to take up two or three references, one of which will be your bank, and they will allow you thirty days credit. Some add a percentage to your statements for overdue accounts and if this is in the terms you agreed at the beginning, you will have to pay it. Some offer small discounts for prompt payment, but this is becoming less common. What most of them do is make a carriage charge for delivery, but waive this if the order value is above a minimum level.

Some manufacturers and wholesalers suggest a recommended retail price (RRP) on certain items. This may be rather more than the usual mark-up and in these cases it is wise to accept that they know what they're doing. Another of the great rules of business is that everything is worth what someone is prepared to pay for it, and that price isn't necessarily related to the cost of production. Most items, however, have no RRP and it is up to you how much you charge. However, now that many retailers (and even manufacturers) are selling by mail order and the internet, it is easy for customers not only to compare prices but also to obtain the goods they want elsewhere if they don't like your prices.

For books the normal discount is 35% whether you buy from a wholesaler or the publishers, but most publishers make a surcharge for small orders and many of the big general publishers no longer supply retail outlets. Wholesalers generally carry large stocks and thus can deliver quickly. If you want a mixed selection of books for non-horsy customers, it might be worth going to a 'remainder' merchant, who will offer packages of books at a very good price. (Remaindered books are those which publishers have decided to abandon, so they sell the remaining copies at very low prices to these remainder merchants.)

Whether you buy your stock from a manufacturer or wholesaler is up to you. The price will be more or less the same but delivery times may be longer from manufacturers and they may have 'minimum quantity' rules. Most of the big wholesalers now carry enormous stocks and reckon to deliver within forty-eight hours. There are some situations (mostly with 'big name' clothing) where you can buy only from the manufacturer, and they may not be prepared to supply you if they already supply someone else in your area.

Both manufacturers and wholesalers employ representatives who call on retailers, to show new lines rather than just take your regular order. It is normal for this to be by appointment. Most of them will carry a fair amount of stock in their cars in case you need something urgently.

Incidentally, if you are toying with the idea of selling cheap tack from your house or at boot fairs, don't bother. You could have got away with it in the days when it was common practice to go to Walsall with a pocketful of cash for your supplies, but those days are long gone. The manufacturers and wholesalers are far too conscious of the need to keep their bigger outlets happy, and you will find it very difficult to get stock.

How do you find all these suppliers? You could peruse the columns of the horse magazines but this won't be very rewarding, as few trade suppliers advertise here. You could place an advertisement in *Horse and Hound* to announce your venture and ask suppliers to contact you. A better place to put this advertisement would be in *Farm & Country Retailer*, or *Equestrian Business News*. Better yet, buy a copy of the *Equestrian Trade News* and its annual *Trade Suppliers Directory*. Best of all, join BETA itself.

## Second-hand Stock

There are, of course, always second-hand items. Many purchasers of new saddles expect you to take their old one in part exchange; other people have a periodic turnout of their tack rooms, or give up riding. If you don't want to get involved in selling part-exchange saddles you can either pass them on to someone who does, or send them to an auction sale.

There are a couple of advantages in stocking second-hand items. One is that people come a long way to see your second-hand stock, then decide they'd rather have new and buy it from you while they are there. The other is that it is normal practice with used goods to sell them for the owner on commission, which means you have no outlay and pay VAT only on the commission. If you actually buy goods to sell, you have to pay VAT on the whole of the sale price, often without an invoice that allows a VAT reduction on the purchase price.

If you do buy second-hand items, naturally you do not pay

very much for them, for they are not as easy to sell as new. Finally, you must be very certain that the items you are buying are not stolen and that you get proper receipts (with addresses) for each item. Whichever way you acquire it, your record-keeping must be as meticulous as it is for new stock.

## Sale of Goods Acts

Whether the goods you are selling are new or used, you are still bound by the Sale of Goods Acts. These are long and complex, but the important bits are as follows. Firstly, goods must be of 'merchantable quality' which basically means they mustn't fall apart. Next, they must be 'fit for the purpose for which they are intended', which is to a certain extent self-explanatory, but includes the fact that a customer is entitled to rely on your 'professional skill and judgement', so if they have asked for your recommendation, you'd better be right.

In practical terms it means that if the goods are faulty or not right for the intended purpose, you take them back and refund the money without question. But it could also mean you being sued for damages if the goods have been the cause of an accident – so check your insurance policy. Incidentally, since *you* sold the goods, *you* carry the can, not the manufacturer or the wholesaler, although, obviously, you can expect them to replace faulty items.

## Repairs

Last, but by no means least, we come to the matter of repairs. I am assured by all wholesalers and retailers I have talked to that it is essential to provide a repair service. You can get by with sending equipment out to a repairer, but the best way is to have a resident saddler. If you can find one who will pay you a small fee for using your premises, rejoice. If not, and you have to employ one, grin and bear it. It will probably not be economic on the face of it, but it will bring the customers in and that is what this whole business is all about.

# 3
# *Stock Control and Money Matters*

## *Fixing Prices*

If you are a newcomer to the saddlery trade, prepare yourself for a shock when you find out the difference between saddlery wholesale and retail prices. In most retail businesses the norm is around 25% mark-up – in other words, if you buy for £1.00, you sell for £1.25. But this applies in shops where the stock turns over ten or twelve times a year. I mentioned earlier that saddlery turns over only two-and-a-half times a year and it is for this reason that the normal mark-up tends to be at least 100% (i.e. buy for £1.00, sell for £2.00).

So with all the items which do not have a manufacturer's recommended price, you work on a basic pricing rule of doubling what you have paid, then adjust it for what the local market will bear. If you are the only tack shop for fifty miles, customers have a choice of paying your prices, making a one hundred mile round trip or paying postage on mail order. They might moan a bit, but as long as you are not too outrageous, they'll pay. But if there is another shop three miles away, your pricing options aren't too wide.

By the time you've paid your rates, electricity bills, insurance premiums and rent or loan interest on your lease (or the loss of interest your own capital could have been earning elsewhere), you have hefty outgoings even before you buy stock. Add another chunk of interest on the stock value and you will realise that even 100% mark-up doesn't give you a lot of leeway for errors of judgement.

The classic error is to give a discount to the local riding club members. 10% is the usual amount and it just about eats away any profit you might have made. *There is absolutely no need to do it.* If your stock is right and the general level of your prices is right and you are conveniently situated, they will buy from you anyway. Unless someone intends to buy in bulk and makes it clear they will only do so if you give a discount, don't. Even in this situation, don't do it unless the sale leaves you enough stock to satisfy your regular customers (at the full price) until you can get a delivery.

There are many astute shoppers around who will want to haggle over prices. Whether you are prepared to accept this has to be dependent on the value of the items in question and the value of that customer's ongoing business. If a stranger offers a reduced amount for something you've been stuck with for a long time, it is sensible to accept rather than leave the item on the shelf, but if the item in question is something that sells easily, why make a gift to a stranger?

The reduced price has to take into consideration not only these factors and what the item cost you, but also the method of payment. What normally happens is that the customer offers cash; sometimes with the suggestion (spoken or implied) that you fiddle the VAT. This is illegal, and since the VAT man knows full well that it happens and is on the lookout for it (and they've heard all the excuses about stolen stock) you'd be extremely foolish to do it. However, if you accept cash for sales, you do not have to suffer the charges of a credit card company, and it's no skin off your nose to give that small percentage as a discount.

## Stock Records

The possibility of haggling over prices means you must have stock records of a sort that allow you to check quickly and easily what you paid for things and when you bought them. Since you will probably have at least two hundred different items in stock, you cannot possibly keep this information in your head accurately. Even if you could, you will still have to keep proper records for your accountant, the VAT man, possibly the Inland Revenue and possibly stock auditors if ever you want to sell the business.

Without proper records, you cannot evaluate the profitability of various items. Nor will you be able to establish a pattern of sales, seasonal or otherwise, to help you with re-ordering. It is pretty obvious that few customers will want fly repellent in January or heavy rugs in July. But if they start buying that fly repellent at the end of March and it has a four week delivery, you need to order your first supplies in February and your last in July. It is not unreasonable to assume that you will sell about the same quantities in each month as last year, but without proper records, how are you going to know what those quantities were?

There's a concept called 'just-in-time' ordering. It originated in the car industry, where holding unnecessary stock could mean a car manufacturer had millions of pounds tied up in parts. It won't cost you that much, but if you have an overdraft or other loan, it costs you the interest on the part of the loan that is tied up in stock, and if you don't have a loan, it costs you the profit that money could have been earning on a faster moving item. With items which have a long delivery time, it is important to know when they sell and in what quantity.

Another not unimportant point about keeping small stocks is that it is easier to spot what the supermarkets call 'leakage' (theft), and this in itself helps to keep sticky fingers away. If you normally only keep three of something on the shelf, you'll soon notice if one goes missing, but you are less likely to notice there are only nineteen when there should be twenty. The point here is that if you can pinpoint *when* things went missing, you have more chance of doing something about it.

You must do a full stock-take at the year-end, so that your stock can be reflected accurately in your accounts. If you have a big shop, lots of stock and several members of staff, you should do it more frequently and insist on all sales being recorded by a stock number as a safeguard against staff theft being disguised as incompetent till handling. Best of all, maintain a constant running stock-check as part of your re-ordering and stock display systems.

Once you get beyond the stage of stocking more than a couple of hundred individual items, you will need a computerised stock system. For a large scale, you need what is known as an EPOS (Electronic Point Of Sale) system, which links the tills to the main back office computer, automatically updating the stock records every time something is sold, and warning you when anything gets to its re-order level. All your stock will have to be marked with bar-codes, the tills will have to be equipped with a scanner, and you will need some hand-held scanners for stock-taking. All this is going to be quite expensive, and I'm told these systems aren't foolproof – the batteries in the scanners fail and other irritating things can go wrong, as you've probably seen in your local supermarket.

The next level is manual recording of large item sales combined with frequent small-item stock checks, with the results being recorded on your computer. Alternatively, you can do the whole thing on a set of index cards.

Whichever way you do it, you need the description and stock number of each item, plus columns for dates, purchases and cost, VAT element of cost and sale price, numbers sold (or missing), number currently in stock, and the supplier's name and catalogue number.

You should certainly have some easily identifiable distinction between items that do not bear VAT and those that do. With an EPOS system it isn't a problem, as the system does it all for you; otherwise colour coding on both stock cards and price tags is the easiest way to do this. You will also have to differentiate between such items in your manual day-to-day book-keeping when you record your receipts, since the VAT inspectors will want to know how you do it. One way would be to keep the price tags from each sale and compare them with your total receipts at the end of the day. Another would be to list each item in one of two columns, as it is sold. However, even the simplest of cash registers should be capable of recording on the paper roll inside the machine whether or not items sold bore VAT. You should keep these as part of your accounting records.

## Value Added Tax

You will almost certainly have to register for VAT. The rule is that if your turnover is above a certain level you must register. The level changes each year with the Budget, but it is so low that you are unlikely to be below it. Note that 'turnover' means takings, not profit, but it does have the proviso that the takings relate to what are called 'taxable' items. Some items are exempt from VAT, others are considered taxable but the rate is currently set at zero, and the rest are taxed at standard rate. Unfortunately, although some kinds of animal foodstuffs and childrens' clothing are zero-rated, they are still taxable and most of your stock will be standard rated.

If in doubt, ask your accountant or your local VAT office. The latter are very helpful, especially when you are starting up. It's only when you are late with your returns, or they think you might be on the fiddle, that they get nasty. And they can get very nasty indeed, for they have the powers of search and seizure of anything they think might be relevant to their enquiries. That means your home as well as your business premises.

Once registered you must keep proper financial records (see Chapter 4), collect VAT from your customers and hand it on to the Commissioners for Customs And Excise. On registration, you will be given your own VAT number, and any invoice you issue must show that number and generally comply with the regulations. You have to submit a quarterly return, showing how much VAT you have collected, how much you paid out on purchases, and send your cheque for the difference. In the quarter when you first open, you may well have paid out more on your new stock than you will have collected in sales and in this case they will reimburse you. The forms are quite simple to fill in, but if you don't want to deal with it yourself, your accountant will do it for you.

## Accountants

No one should try to operate a business without a good accountant. Not only do they know all about taxes, and provide a convenient buffer between you and the Inland Revenue, they will also perform a multitude of other services. They will do your book-keeping, work out wages and PAYE – almost anything you can think of. But do remember that they charge on time spent, so expecting them to work from scrappy bits of paper, having to telephone you for clarification of details, and other time wasting occupations, is going to cost you extra.

As a general guide, it is good to choose an affluent accountant. If she can't handle her own financial affairs properly, she won't do much for yours either. The one you want is going to suggest to you how you can save paying tax, not wait until you ask her how to do it, as well as doing straightforward accounting work. How do you find the paragon you need? Ideally by personal recommendation from someone who is in the same line of business as yourself which, in this context, includes other types of shop. If that fails, ask your bank manager to recommend one. Indeed, most banks have a department of their own which does this sort of thing, but remember that if you are trading as a limited company, your accounts must be audited by a firm of Chartered Accountants.

## Bank Accounts and Overdrafts

You really must have a separate bank account for the shop, even if you are running it as a sole trader. Although your accountant could cope with you doing everything from your own personal account, it means more work for her and more cost for you, as above. The Inland Revenue won't be keen on it, the VATman won't be keen on it and nor will any prospective purchasers if you want to sell the business.

It is also a good idea to have an overdraft facility. This means that if you need to overdraw, you can always do so up to an agreed level, not that you should have a permanently over-drawn balance. The trick with an overdraft is to keep it as low as you can by banking your takings every day and by careful timing on paying your bills, for although the rate of interest will be a couple of points higher than for a formal loan, you only pay it on the amount outstanding on a day-to-day basis.

# 4
# *Financial Record Keeping*

Whatever you give your accountant at the end of the year in the way of accounts producing material, you must keep day-to-day records yourself. Remember that the VAT inspector may call and demand to see your records and woe betide you if you are more than a month behind.

## *Computer Systems*

Assuming that you have basic computer literacy (and if you don't, you can soon acquire it at adult education classes) the best way to keep records is on a computer. You don't have to be a programming whiz, as there are many accounting packages designed for small businesses (e.g. Sage, MYOB, Quicken, etc.). You will need a package which allows you to keep records of your bank accounts, stock, accounts with suppliers, perhaps accounts with a few big customers, do your staff wages and VAT returns and produce cash flow forecasts. The joy of such packages is that they automatically perform calculations such as adding up columns of figures and calculating PAYE. A good stock control system will also warn you when you need to reorder.

Of course, if you are a computer whiz, you could create your own programs, but quite frankly the aggravation involved (think of inputting all those PAYE codes) isn't worth it when you consider how reasonably priced the packages are. Equally, it would not be worth paying a programmer to write a suite of programs for you. So far as I am aware, there aren't any special programs written for tack shops but any package that is suitable for retail shops will do the job. For this purpose, there isn't any difference between salt licks and soap powder!

Before buying a package, check with your accountant as it should be possible to give her a disk at the end of the year and thus reduce her fee.

## Manual Systems

If you don't have staff PAYE to calculate and you only offer a small range of stock, you can keep manual records. The most basic method is to do everything in one book called a cash book where you list everything without attempting to analyse it. Get a ruled book from the stationers, use the left hand page of each pair for income and the right hand page for outgoings. This is the way every other business does it, so don't confuse your accountant's staff by being different. Make a note each day of how much you have paid into the bank, how much you have paid out and whether it was cash, cheque or credit card. Keep all invoices and receipts. If you have taken cash out of the till for any reason, make a note of that as well.

This job is going to take you about thirty minutes a week and it is the minimum you must do. It is going to take your accountant a long time to unravel it and check it against your bank statements and she will charge you accordingly. It is a method which does not give you much information to help you with your planning, either.

The other way to keep books is only a little more time-consuming for you, but it saves your accountant hours of tedious work and thus keeps the fees down. It also provides you with at-a-glance information on your financial affairs. For this, you'll need a little more in the way of books. You'll need one book for petty cash (called petty cash book) and one for bank transactions (this one is called the cash book). Both will need to be ruled in cash columns, with a wide column on the left for details and dates. The petty cash book need not be large as it will need only about six columns per page, but the cash book will need to have several columns each for income and expenditure. You can buy these books ready ruled at any stationers. It does no harm to have two sets, each of which you use in alternate years, while your accountant deals with the other at her leisure without depriving you of your working books.

For the cash book, starting a new page for each month, head up the columns for income and regular expenditure, allowing one column for the VAT element in each case. Keep the first column on each side for basic bank details.

The entries on the expenditure side should consist of cheques paid out, any standing orders you may have and charges made by the bank. You should have a column for each type of regular expenditure, such as rent, wages, utility bills and stock purchases. You might want to separate stock into categories such as saddlery, clothing, veterinary items, etc. The entries on the income side should correspond with the daily totals from your cash register. You should bank your takings every day, especially if you have an overdraft. You may wish to separate your takings into cash, cheques and credit cards.

When you want to pay some money into the bank, draw a line under the last entry in the first column, add the amounts and put the total in the second column alongside that line. This amount should tally with the total on the paying-in slip and this makes it much easier to check off bank statements. For expenditure it is equally easy. When you write a cheque, fill in the stub with not only the date and amount of money, but the name of the person you paid and the item you paid for – and if VAT is involved, the amount of that. (Example: 12th July 1999. Bloggs & Co., wholesale. Invoice No. 1498. VAT £175, Goods £1,000, total £1,175.) Then all you have to do at cash book time is put down the name, date and cheque number (so it can be checked against the bank statement), the amount of the cheque in the first column and allocate the amount as before into its respective columns.

For expenses, there will be items that crop up occasionally, but not often enough to have a column of their own, like your shop insurance premium or the three-monthly payment to the VATman. For this purpose you have a column marked 'Other', with a space next to it where you can note what it was.

For the petty cash book you do not need so much detail, as the only income will be when you cash a cheque for this purpose. The expenditure will be such items as coffee for the morning break, a tip for the dustman or stamps from the Post Office.

Doing your book-keeping this way will take an hour a week and another hour at the end of the month, when you add up all the columns and put the invoices in order. Put a paper clip on each month's collection and put them away safely. With these columns added, you can not only see how you are doing, but your VAT return will be easier and you will have figures handy for your cash flow.

To recap:
- Use left hand side for income, right hand side for expenditure.
- Fill in cheque stubs properly.
- Complete paying-in slips properly.
- Keep all invoices, bills, receipts, till rolls, etc. in order.
- Start a new pair of pages for each month.
- Add up each column at the end of the month.

And the golden rules:
- Note petty cash expenditure and details of takings daily.
- Bank your takings daily.
- Ensure your bankings agree with each day's till records.

*Examples of both cash and petty cash books are shown in Appendices 1 & 2.

## Budgeting and Cash Flow Forecasts

One task your accountant will press you to attend to is budgeting, which is a fancy term to describe the simple fact that you should be prepared to meet certain expenses at regular intervals and must have the money available. Consider the implications. Many of your expenses, such as wages or stock bills will have to be paid every week or month, but others, such as telephone, rates or loan interest are only due at three- or even six-monthly intervals. You need to make sure the money for these is available without having to exceed your overdraft limit.

Budgeting is a task that causes even the most experienced of business people to quake in their shoes, which is silly. Like many jobs that you put off doing because you think they are going to be awful, when you actually get down to it, budgeting is easy and quick to do. You know exactly when and what your loan repayments (or rent) and wages will be, and you can make an educated guess at the electricity bill and stock purchases and earnings from sales. All you have to do is build this information up into an easily readable form which is known as a cash flow forecast.

To do a cash flow forecast on paper you'll need two large sheets of ruled and columned paper: you will need one wide column on the left for details and eighteen small ones for figures. On the first sheet, taking two columns per month, head up the first twelve columns for the next six months. The next four columns will be for the last two quarters of the year and the final two for totals. Assuming that you are starting at the beginning of the year, you should now have pairs of columns marked January, February, March, April, May, June; July to September and October to December. Head up the second sheet in quarter years and totals. That gives you the next two years. Finally, mark each pair of columns 'Actual' and 'Budget'.

Now for the details. Use the top of the page for income and the rest for expenditure. Allow three lines at the bottom for totals, balances and cumulative balances. Allow one line for each item you spend on (rent, loan repayments, rates, wages, stock, etc.), and make sure you have one marked 'Contingencies'. This is to allow for all the disasters that can hit you when you least expect them. If you allow some money to deal with them, they won't hurt so much. When it comes to expenses, don't forget to include your own wages.

Having done that, you can begin to put in some figures.

Work out how much each item earns/costs each month and fill in the budget column – and don't forget seasonal variations like winter rug sales and pre-Christmas gift sales. This will be easy for the first six months as you can be fairly sure what things cost, but you must make allowances for rising costs further ahead. I would suggest a 5% increase every six months.

Having done this – and you must make allowances for *everything*, you can then add up the columns. First total the income and expenses, then deduct expenses from income and put the result in the balance line. If expenses exceed income and you get a minus figure, put brackets round it. (This is an accounting convention to indicate a minus figure.) Now you can do the cumulative balances. Just add the month's balance to the previous month's cumulative balance. Hopefully, this will increase steadily as the months progress, but it may go into negative figures at times.

You now have an advance picture of the development or decline of your business. It may scare the living daylights out of you, but it will tell you the truth about your chances of success or bankruptcy. It will prepare you for the lean months, so you can have funds ready, instead of that panic to sell something below its real value to pay the bills. It will tell you when you will be able to afford a new van, or if you can afford the repayments on a loan to build the extension you desperately need. It will also convince your bank manager not only that you can afford these repayments, but that you are a responsible business person instead of a feckless fool, blundering along blindly in the dark, hoping everything will turn out right.

Preparing regular cash flow forecasts on a computer is much easier, because once you have a spreadsheet set up, it will automatically calculate the totals, and if you need to change any figures, it will automatically change the totals for that column and those which follow.

One final point about these cash flow forecasts. The more you do them, the better your guesstimates of amounts will become, and this is where those 'Actual' columns come in. Every month, when you add up the columns in your cash book, fill in the 'Actual' column, on your cash flow. Then you can compare Actual and Budget and see how accurate your guesstimates were – and demonstrate that accuracy to the bank manager.

*An example of a cash flow forecast is show in Appendix 3.

# 5
# *Profits, Losses and the Taxman*

Everyone knows what a profit is. It is the difference between what something costs you and what you sell it for. And if you sell it for less than it cost you, then the result is a loss, not a profit. But there is a snag, as always. Actually the profit I just mentioned is called gross profit and it is subject to tax. When the taxman has had his bit, what is left is called net profit, and that is what you have to live on. So obviously, the less the taxman gets, the more you get and so we come back to the point I keep labouring, which is that a good accountant is worth her weight in gold, for her tax know-how will save you more than her fee.

## *Taxation*

Taxation is a complex field, and since each national Budget tends to change the situation, the help of an expert to maximise opportunities is essential. All you have to understand is the difference between tax evasion and tax avoidance. Evasion is illegal and dishonest, as it involves deliberately taking action to evade paying tax that is lawfully due. Avoidance is perfectly legal and it hinges on a famous court decision that said that there is no legal bar to arranging one's financial affairs in such a way as to reduce one's tax liability to a minimum.

The main area in which a small business can reduce its tax liability is by ensuring that all legitimate expenses have been claimed. A legitimate expense is one that relates to the running of the business. Coffee for the proprietor's home use is not, but coffee for the staff's morning break is, as well as the cost of the chairs that they use while they drink it. Dog food might be, but not if you live a long way from the shop and take the dog home with you. There is no laid-down rule as to the type of dog, the principle here is that it is a guard dog, which means it has to be on the premises it is meant to be guarding. By all means take it to shows with you, for then it will be guarding your stand or van of stock.

What all this boils down to is that you must keep the bills for *all* expenditure and let your accountant decide what is legitimately claimable. And if you are considering making any radical changes in your field of operations, buying any expensive equipment, or even changing your car, then consult her first. It might be advantageous to time these moves according to your year end, the last Budget or just your general situation. Trust your accountant and let her use the financial skills you are paying for, to your best advantage.

## Limited Companies

You may feel that you should start a limited company for your business. Whether or not this is a good idea depends, in the final analysis, on your individual circumstances, and you must discuss these with your accountant, but as always, there are pros and cons.

Firstly, it costs more to set up a limited company than to start trading on your own account (called sole trader) or as a partnership. There are stamp duties to be paid, and certain statutory books must be purchased, as well as a company seal. The usual procedure is for your accountant to purchase a ready-made company*, complete with everything needed, from a firm which specialises in starting companies. This is quite cheap so long as you are prepared to accept a neutral

*This will cost about £100.

name, but if you want a specific name it will cost more. Your accountant will then charge you for preparing lists of shareholders and directors which must be sent to the Registrar of Companies. To commence trading as a sole trader or partnership, you do not even have to register a business name, even if you are not using your own name but something like 'Bestprice Tack Shop'.

Next, a limited company has to submit a list of shareholders, directors and other details, called an annual return (complete with a copy of the audited accounts), to the Registrar of Companies each year. You must pay a small fee every year for the privilege of having these documents filed at Companies House, where anybody can go along and look at them to see what you're doing. Sole trader and partnership accounts do not have to be filed anywhere, and you may prefer to maintain your privacy.

Limited companies are subject to Corporation Tax on their profits before dividends are paid – which effectively means that you will be taxed twice, as you will also have to pay tax on the dividends when you receive them. This last point may be purely academic, as you should be drawing all profits as director's salary or putting them into your pension fund. More on that later.

There are also certain legal restrictions on exactly what you may or may not do with a limited company and its money, which do not arise with a sole trader or partnership. These may seem a lot of points against having a limited company. They are, however, all offset by the great advantage of having limited liability.

If you are a sole trader and your business fails, your creditors can make you bankrupt and that means that you can lose everything you own, apart from your clothes and the tools of your trade. If you are a member of a partnership, you may also be liable for the debts of your partners (*all* their debts, not just those relating to the business you are involved in!), unless you get a solicitor to draw up a formal Partnership Agreement which states that you do not accept that liability.

If your business is run by a limited company, that company

has a separate legal entity and if that entity goes bankrupt, the property of its owners cannot be seized to pay off its debts. You may lose everything the company owns, but you do not lose your house, your car or the contents of your personal bank account. It is for that reason that many small businesses are content to put up with the expense and inconvenience of running a limited company. However, you will find that bank managers are cagey about lending large sums of money to limited companies which have just started up, and they will probably require your personal guarantee on the loan, which puts you back in the position of losing your personal property if the business goes bust.

One further thought on bankruptcy. A creditor can start bankruptcy proceedings against you if you owe him as little as £750 and fail to pay. And what is not generally known is that at least half of the country's bankruptcies are brought by either the Customs & Excise against defaulting VAT payers, or by the DSS against people who do not pay for national insurance stamps. This last is very easy for the small business to forget, but they always catch up with you in the end!

One point on running your business as a limited company which needs careful consideration, is whether you personally want to be employed or self-employed. If you are a sole trader you will be self-employed and get all the tax advantages which attach to that. But you will get very little in the way of social security and unemployment benefits, and for this reason you may prefer to be employed, which is easier to organise where a company is involved. There are also some major advantages attached to your pension arrangements. This is a very complex area which you must discuss in depth with your accountant before making a decision. More on pensions in Chapter 6.

## Losses

Finally, some thoughts on losses. From an accountant's point of view, a loss is not always the disaster it may appear to the layperson. If it goes on too long, the taxman will begin to wonder how you manage to eat regularly and start to probe deeper, but on a short-term basis (like about four years) it may even turn out to be advantageous. The Inland Revenue get a bit tired of clever business people using their spouse's hobbies as tax losses, but provided you can prove that you are running your business with 'a view to profit', they will not object too much. But beware those words and if you get a letter from the Inland Revenue using them, be sure to consult your accountant before replying. Come to think of it, it is a good idea to consult her every time before talking or writing to them. They have a sneaky habit of asking superficially innocent but actually carefully worded questions that need equally carefully worded answers.

How can a loss be turned to advantage? Losses from one year can be set against profits from the next and this reduces the tax on the profitable year. Remember that as a self-employed person you are entitled to the same personal allowance as the employed person, which means you pay no tax on the first chunk of your income. Suppose that allowance is £5,000 and your business shows a profit of £5,000 plus £600 after last year's losses have been deducted, you will only have to pay tax on £600.

Better yet, losses from one business can often be set against earned income from another source, with the same result. So, if you are running a business on the side and you have another job, or if a married couple are taxed jointly, one works and the other runs the shop which makes a loss, then that loss can be used to reduce the tax liability from the job. And since the tax on the job will already have been paid by PAYE, the end result is a tax rebate.

# 6

# *Insurances – Obligatory and Sensible*

## *Obligatory Insurances*

Everyone knows that all vehicles must carry third party insurance, but it is not so generally known that businesses must also carry certain insurances to protect their staff and the public. These obligatory insurances are as follows.

### EMPLOYER'S LIABILITY

This insurance covers such items as unsafe working conditions or equipment, accidents to staff while they are working, or injury caused to them through your negligence. Not only are you obliged to carry this insurance, but you must prove it by displaying a certificate in a prominent position. Failure to do this could result in the authorities closing your establishment.

### PUBLIC LIABILITY

This insurance covers you for injuries to members of the public both while they are on your premises and when they are passing your premises. Such injuries might arise from bits falling off your buildings, or from obstructions such as wheelbarrow handles and stray tools (shovels, forks etc.). It also covers you what for insurance companies call 'product liability'. This is the situation I mentioned earlier where an item you recommended has proved faulty and caused an accident, and the even nastier situation where some repair work your staff have carried out has failed and caused an accident, or even where you have acted as an agent and done no more than pass repair work to an outside repairer.

Considering the magnitude of some recent awards for injuries leading to incapacity, you should be adequately insured for both these possibilities. Cover of £2 million is not excessive. Be sure your policy covers not only your shop, but also your show stand if you have one, and such situations as you or your saddler going to a customer's premises for fittings and deliveries.

Finally, keep an accident book and enter *all* incidents, however trivial they may seem at the time. Even a small cut on a finger could lead to blood poisoning if neglected, and you could have forgotten all about it months before a claim is made. If you can prove, from your accident book, that you cleaned and disinfected that cut and put a sterile dressing on it, you will save a good deal of hassle and expense. Just list the date and time, name of the victim, cause of the injury, nature of the injury and the action taken. In the event of a fatality or 'major' injury (one which puts the victim in hospital for twenty-four hours), you must also report it to the Environmental Health Department. If you are unfortunate enough to have a death on your premises, you must immediately report it to the police.

## Sensible Insurances

### FULLY COMPREHENSIVE VEHICLE COVER

Driving around in a damaged vehicle does nothing for your image and you may not be able to afford to have it repaired without insurance cover. Nor may you be able to afford to replace stock that was stolen from it, or with it. Just be sure to tell the insurance company you are using the vehicle for your business, or they may decline to pay out. The stock itself will need 'stock in transit' cover. Because these and other important insurance policies tend to be interlinked, it is easiest and best to do all your insuring through one broker.

PREMISES

The next item is your premises themselves – the actual bricks and mortar known as the 'fabric'. The amount of cover should be for the value of the building itself plus 15% for clearing the remains from the site in the event of total disaster. Even if you only rent the premises, you may still be obliged to put this cover in place, so check your tenancy agreement. If you have a lease, the freeholder will insist on it; so will the bank if you are using the lease as security for a loan.

CONSEQUENTIAL LOSS

Cover is also available for 'consequential' loss. This only comes into play when disaster strikes your premises and prevents you from trading. It may not give you *all* your lost potential profit, but it will provide for fixed expenses such as your rent, wages, etc., which have to be paid even if you are not trading.

STOCK AND ITEMS UNDER REPAIR

Your stock is covered for its wholesale value. You pay a premium based on the average stock value which will be somewhere around your stock-take figure. The fact that it is saddlery is irrelevant. The only time that extra premiums are involved is when a shop's stock includes what the insurance companies call 'high resale' items like cigarettes or alcohol.

Stock cover does not include such items as customers' property in for repair, or to be sold on a commission basis. These items would be listed in a schedule headed 'customers' goods for sale or repair'.

Shop fittings are usually considered to be part of the stock. They will include cash registers and safes, glass counters and mirrors but not necessarily plate glass windows. These may be covered under 'fabric', but you may have to cover them separately. Whichever, it will include a twenty-four hour boarding-up service.

Most general insurance companies offer all or most of these insurances (except motor vehicle cover) in a comprehensive 'shopkeepers' package and you select those parts of the package you want. Rates for everything except the fabric element vary according to your location. If you are situated in a village, you will pay only half as much as you would in a small town or a third as much as you would in Greater London. Rates for fabric, consequential loss, stock and fittings may be as much as four times higher for a timber building as for one of conventional construction.

### COMPUTERS

Finally, if you have a computer, you will need to insure this as well, whether it is on your business premises or at home. Laptops are a favourite target for thieves as they are so portable and so easily sold, but even desktop PCs are vulnerable. The tendency these days is for thieves to remove the chips rather than take the bulky processor, monitor and printer, but even that will cause a major disruption, especially if you don't make regular backups.

## Under-insuring

Another warning on all these items. If you under-insure (e.g. you only cover half your value) you may find the insurance company does a nasty thing called 'averaging' when you make a claim. It works like this - stock worth £50,000, you insure for only £25,000. Half your stock is stolen, but they pay you only £12,500 because they say you are accepting half the risk yourself! The only way round this is to cover specific items only - that £25,000 might be the value of your saddlery stock and not include clothing. In that case, the insurance company will pay out the full £25,000 if both saddlery and clothing are stolen.

## Personal Insurances

Now to the personal side of insurance - the policies that relate to you and your family. I mentioned in Chapter 1 that you will be required to insure your own life if you have a large loan.

There are a number of ways you can do this. The cheapest is called 'term' which pays out if you die within the fixed term of the policy, say five years. These policies have no surrender value while you live since the whole of the premium is used to purchase life cover.

Slightly more expensive is 'whole of life', which pays out on death whenever it occurs, but unlike 'term' will have a surrender value. The most expensive is 'endowment assurance' which is written for a fixed term, at the end of which it pays you a lump sum. Part of the premium, as in 'whole of life', is used to buy life cover and the greater part is used to buy the investment element necessary to guarantee the payout at the end of the fixed period. This type of policy pays out the death benefit within the fixed period, but also has a surrender value.

Should you be unlucky enough to fall sick for any length of time it is likely that your business will suffer permanently, especially if you have no money coming in while you are unable to work. For this reason you may want to take out a 'permanent health' policy, which pays out while you remain unable to work, until the age of sixty-five. These are not easy policies to get because the insurance companies are wary of them for the very reason you want them! They cannot cancel the contract to insure you until retirement age no matter how often you make a claim. This is most important if you are self-employed, for you will get little sickness benefit. You might be able to get someone to take over running your business while you are ill, but will it make enough to pay both them and you?

The snag with permanent health policies is that they do not start to pay out for the first period of sickness – usually thirteen weeks. For this reason you should also have a 'sickness and accident' policy which can, according to which company you buy from, pay out a monthly sum for as long as two years from the start of incapacity.

You may also feel it is wise to join one of the private medical insurance schemes, such as BUPA or PPP. These ensure you prompt treatment at a time convenient to yourself, should you require an operation. Imagine trying to run your shop while feeling increasingly ill for the year or more it can take to get treatment on the National Health. You may survive, but will the business?

## Personal Pensions

Don't make the classic mistake of assuming that you will be able to sell the business when you retire and live off the proceeds. You might be able to sell – but all it will fetch is the wholesale value of the stock, a 'premium' as mentioned before, and whatever remains of the value of the lease, which altogether rarely adds up to a sufficiently large sum to keep you in luxury for the rest of your days. But what if it is not saleable because of a new motorway, or because urban sprawl has driven all the horse owners away, or because the lease is about to expire anyway, or because your failing health has run the business down to such a level that no one wants it? It would be sensible to make the proper provision for your retirement, especially if you can do it instead of paying lots of tax!

If you are operating on a self-employed basis, you will be able to take out a 'personal pension plan'. The tax advantages of these plans are enormous, for you effectively give yourself a substantial chunk of what you would otherwise have given to the taxman. You can put a big percentage of what the insurance companies call 'net relevant earnings' (in other words, your business's gross profit after deduction of expenses but before tax) into your pension plan each year, and the older you are, the bigger that percentage becomes.

If you are operating as a limited company and taxed on a PAYE basis, your pension will be not a 'personal' but an 'executive' plan. Here you can contribute considerably more than above, to such an extent that you can virtually eliminate any tax liability perfectly legally. As you might imagine, with such advantages, these schemes are extremely complex and you will need to consult both your accountant and a financial adviser who is more experienced with this type of insurance plan.

With both types of pension, you have the options at retirement of taking a pension, or leaving it to grow a little longer. And you don't have to make the decision on what to do until retirement. Both have death benefits and a facility for a widow/ers pension after you die.

There is another advantage of paying into a pension plan, which is that you can use it to raise a loan. The security is the lump sum at retirement and you can borrow up to fifteen times the annual contribution. These loans can be cheaper than other types, but once again you will have to consult an expert.

Where pensions and life assurance are concerned, you can get advice from two types of source. The first is from an organisation such as a bank or building society, which sells only its own financial products and can thus only advise on these. The other is an independent financial adviser (known as an IFA) who is able to sell and advise on products from the whole of the financial industry. Both are now rigidly controlled under the Financial Services Act*.

One final word on general insurances. Shop around a little before you buy. Ask friends which company of brokers they have found most reliable – for remember, brokers exist on the commission they get from the insurance companies, and this may influence their advice to you!

---

*The comments in this book on investments and investment-linked products are intended for information only, and are not intended to constitute investment advice.

# 7
# *Staff*

There will be many occasions when you will have to leave your shop, even if you do not actually have to go out to visit customers or collect from suppliers. You will want to have a holiday, may have to take time off for sickness and you'll need a lunch break and to go to the toilet. So, unless you have a member of your family handy to help you, you will have to employ someone.

## *Level of Knowledge Required*

It is a big step to have to pay wages and take on the added responsibility of an employee. Not least of all the attendant worries is that of finding the right person. Tack shops are not places where customers just pick what they want off the shelves and pay for it, they are places where the customer asks for help.

While I was researching for this book, I visited a saddler friend to discuss various aspects and I was in his shop for no more than forty minutes, talking to him in between customers. It was a Saturday morning, and of the five customers he served in that time, one came to collect a repair, one asked for a specific item and the other three asked 'What is the best thing for this situation?' In each case, the customer bought something because my friend knows his business. He knows what each piece of tack is for, he knows the competitive situation (one question was 'What sort of jacket do I need for this class?') and he knows his veterinary side as well.

Whoever you employ, if they are to be of any use at all, they must know a fair amount about horses and riding. For this reason you are unlikely to find them at the local Job Centre, nor is a youngster on one of the government-sponsored schemes likely to be suitable. Your best bet is to put an advertisement in your local free horse magazine, if you have one, or in the Riding Club newsletter; on the notice boards at the local riding schools or, best of all, in your own window.

## *Disposition and Appearance*

Apart from know-how, you need someone with a sunny disposition, for there is nothing more off-putting than surly shop assistants. A smile and a friendly greeting do more to bring customers back again than bargain prices.

A smart appearance helps too, so although jeans are perfectly acceptable in a tack shop, do insist that they are clean and topped off by a neat shirt or jumper. The best thing, if you have a lot of staff, is to provide them with jerseys and T-shirts with your shop's name on them. Jewellery should be restricted as it, and loose garments, tend to catch on various portions of your stock (notably curb-chain hooks) with disastrous results. The other thing one has to consider these days is that an increasing number of people are non-smokers and they find the smell of cigarettes offensive. It lingers on clothing, and you don't want your rails of uncovered jackets to smell of tobacco smoke. You are entitled to stipulate that staff do not smoke in the shop, nor in the stock room if you have one, because of the fire risk. (It is not unreasonable to make the same stipulation for customers, since they are even more likely to leave burn marks on your stock. This is best done by displaying a large 'No Smoking' notice on the door.)

## Employment Legislation

For various reasons connected with current employment law, you might be wise to consider employing on a part-time basis rather than full-time. Middle-aged ladies are the best of all, for not only do they tend to have their own transport (essential if you are situated in a remote village), they are usually sensible and reliable. They are unlikely to get pregnant, involving you in the legislation relating to maternity leave, nor will they get together in corners and giggle about their boyfriends.

The legislation is generally referred to as 'Employment Protection' but it encompasses a number of different Acts. However, much of it does not apply to businesses which only employ part-timers or family members, or which have only a few members of staff. This legislation tends to change regularly, and you do need to keep up to date on it. When you are starting up, your solicitor or accountant will be able to tell you which bits apply to your specific situation, or you will find a book on it in your local library. The best way to keep up to date is to subscribe to one of the services such as that provided by Croner (see Appendix 6 for details) which sends you all you need to know as soon as the rules change.

The main provision of employment legislation is that you must give a contract of employment to each member of staff and that you have to be careful about sacking anyone after they have been with you for twenty-six weeks. The basics are these: you must list conditions of employment, hours to be worked, holiday entitlements, wages and periods of notice for ceasing the employment. The usual method of dealing with this is to set it out in a letter with two copies, requiring the employee to sign one copy as agreed and return it to you for your files.

Employees who consider that they have been unfairly dismissed can take you to an industrial tribunal, where you may be ordered to reinstate them or pay them compensation. Even if the tribunal decides in your favour, it will take a great deal of your time as well as horrendous legal costs.

Employment legislation also covers employees' rights to belong to a trade union or professional association; and to maternity leave, pay while on such leave and the right to return to their employment after confinement.

The Equal Pay Act says that employees doing similar work must be paid equally, regardless of sex. The Race Relations Act says you must not discriminate because of race or colour; and the Sex Discrimination Act says you may not discriminate between male and female when recruiting, training or promoting employees unless there is a 'genuine occupational qualification' (such as requiring a male employee to assist male customers in the fitting room).

The Health and Safety at Work Act says that employers have a duty to provide and maintain safe machinery and systems of work for their employees (e.g. leather-working machinery), to provide safe storage and handling of any substance (e.g. wormers), to provide safety information, training and supervision; to take proper care of the working environment and to provide adequate and safe access to and egress from working areas. All these provisions apply equally to non-employed visitors (e.g. delivery drivers), and there are Health and Safety at Work inspectors who have the power to enter your premises and ensure that you are complying with the Act. Obviously, most of these provisions were laid down to protect factory workers, but as you will see from the examples, they can apply to tack shops as well.

You will also have to comply with the provisions of the Offices, Shops and Railway Premises Act. These relate to working conditions for employees. The main provisions of this Shops Act relate to the hours a person may work without a break, as well as the length of that break. The OS & RP Act also lays down conditions on cleanliness, reasonable temperatures (basically a minimum of 16°C or 60°F and you must provide a thermometer in a conspicuous place), ventilation, lighting, toilets and washing facilities, seating, accommodation for clothing, fire precautions and first-aid facilities – and many other conditions. Failure to comply with these Acts can lead to an inspector closing your premises.

If your head is reeling from all this, and its potentially expensive implications, you will probably be relieved to know that you can get insurance cover to help you with legal costs and awards against you. Called 'employer's protection' cover, it is available from most insurance brokers.

## Wages

With all employed staff, you must operate the PAYE scheme when calculating wages. The local Inland Revenue office will supply you with all the necessary documentation and show you how to operate it. Basically it consists of deduction tables for tax and social security contributions, and a record card for each employee. Doing this is another of those tasks that can be frightening in the idea of its complexity, but simple when you actually do it. You can get some kits from stationers if you want to do it manually, or buy a computer program to do it. Alternatively, your accountant will do it for you.

Whatever you do about this side of it, wages should be paid on a regular basis – the same day each week or month – if staff are not to become discontented.

You have to pay sick-pay to employees for up to eight weeks sickness in a year. As always, there are exceptions (and a complex formula for working out the payments) but the main exception is that of employees who earn less than the amount at which they have to pay National Insurance.

## National Insurance Contributions

As mentioned above, you can avoid much of this hassle by employing only members of your family. What you cannot avoid, for they apply to you and family members as well as employees, is National Insurance contributions. Anyone who works and whose remuneration is above a certain level must pay these contributions. In the case of employees, the business is obliged to deduct these contributions from their wages and, after adding the employer's contribution, must remit these sums regularly to the DSS, via the Income Tax offices. Income tax must also be deducted from the wages, and these two amounts added together are paid in one cheque to the Inland Revenue. If you are personally self-employed, you are not exempt from this, you just make a different class of contribution. The easiest way to do this is by standing order.

## The Staff – Boss Relationship

Being the boss is often thought to be a 'cushy' number – all you have to do is tell other people what to do and rake in the money. But those who see it that way forget the other side of the picture. You cannot tell someone else what to do unless you know how to do it yourself, or how will you know it has been done right? The person who is doing the work has only a little section to deal with, while you must oversee the whole thing and accept the ultimate responsibility. You are the one who must ensure that the customers are kept happy, that the right goods are there for them to buy, or there will be no money to be raked in. And yours will be the name on the loan document that provided the capital to make the whole thing possible, so you are the person who must shovel back out the money you have just raked in!

If you find your staff are under the impression that they are earning your living for you, you've been going about it the wrong way. The old 'I'm in charge, don't ask questions, just get on with it' routine is as outmoded as the crinoline. Today's approach is the more effective one of teamwork, with the boss as the leader of the team. Staff are best motivated by a feeling of involvement and being valued, so discuss your plans with them, tell them why you do various things and don't forget to let them know when you feel they've done a good job.

# 8
# *Security*

## *Theft*

As soon as you open any sort of shop you will be considered fair game, both for the criminal fraternity and by opportunist customers. So, high on your list of priorities, even before you order stock, is to make plans for thwarting them. Insurance premiums used to be lower in country areas because the risk was lower but now criminals abound wherever you are and the sort who break in at night like nothing better than premises in an isolated location where it won't matter if they make plenty of noise. It is not unknown for the back wall of a shop to be demolished to make it easier to strip the contents! If you are wondering what the thieves do with their haul, the answer is that a lot of it will be sold at boot fairs and Sunday markets where there are plenty of people who find it convenient to believe that brand new items (even saddles) are 'unwanted gifts' if the price is right.

Another real problem is the innocent-seeming person who comes into your shop during the day. Forget all that nonsense about emotionally disturbed menopausal women who shop-lift on the spur of the moment – shop-lifting is THEFT and it is deliberate. Most of the perpetrators come into your shop with the sole and express intention of stealing something. Watch strangers carefully, but don't trust people you know either. Keep your ear to the local grapevine – the light-fingered are generally known. There was a notorious woman in our district who moved from livery yard to livery yard, leaving a trail of pilfered tack rooms behind her. She was in our yard for a while and we all lost things – and she frequently turned up with brand new items and wanted people to 'ride them in' for her. I often wondered which shop they came from.

Some people have the oddest attitudes. They would be mortally offended if anyone said they were dishonest, but their consciences allow them to put items in their pockets on the basis of 'He's made a fat profit out of me on that saddle/rug/ jacket – I'll have a bit of it back with this pair of gloves/bit/ what-have-you'. There isn't a lot you can do about this, except perhaps send them a bill if you saw it happen.

What you must understand is the difference between the honest mentality and the criminal mentality. When an honest person sees a vulnerable situation, say a purse left on the counter unattended, their reaction is to say 'Whose purse is this? You shouldn't leave it around like that, someone might nick it!' The criminally inclined think, 'I could have that – can I get away with it?' So you must make sure they can't, by keeping all the vulnerable items as inaccessible as possible. I don't mean cheap and easily pocketable items like hoof-picks and mane-combs, but valuable things like a double bridle set made up for display, clippers (a prime target, so keep a note of the serial numbers) and even saddles. People really do steal saddles – honest! I remember being told of one that went from the launch party of a big new shop in London. The owner said 'I saw it go – a man walked in the door, went straight to it, picked it up and walked out with it. There was nothing I could do about it – there were so many people in the shop I couldn't get to him.'

The nastiest type of theft is by your staff. They can develop a 'He can afford it and he doesn't pay me very much' attitude. They might take items home with them to sell to friends, but they are more likely to fiddle what they ring up on the till. (It helps if you have the sort of machine which shows the customer what has been rung up.) Or to have a couple of quid out of the till each day and try to disguise it with over- and under-ringings. Anyone who consistently makes a mess of till-handling is probably doing it deliberately, so beware!

(On reflection, perhaps this is not quite so nasty as the partner who rooks you on the basis 'It's half mine anyway'. Sure, but what about your half, and your half of the profit? One common sneaky deed is to take an item in part exchange, then sell it privately, pocketing all the proceeds. But if you can't trust your partner or your staff or your regular customers, who can you trust? Alas, the answer is no one but yourself – one of the unpalatable but true rules of business.)

## Insurance and Security Systems

All of this means you cannot avoid insuring your stock and premises. Your insurance company will let you know what sort of precautions they want you to take against thieves. This might involve shutters or grilles on the front window and door, and will certainly include locks on external windows and doors and on display cabinets containing particularly valuable items, as well as a safe for cash and a burglar alarm. What they like best of all is an alarm system which activates a signal light or bell in a central control station as well as making a noise on your premises.

Most alarm systems now are electronic and respond to noise, vibration, movement or body heat, so they do not need the elaborate wiring of the older types and are therefore much easier and cheaper to install. They run off batteries or mains (or a combination) and any tampering with the control-box sets them off. They also have built-in delays enabling you to get out of the shop after switching on. Normal practice is to have at least one visible alarm box (or dummy!) situated outside where it will deter passing opportunists. You will have to pay an annual maintenance charge as well as an installation fee although you may be able to purchase the complete system outright and install it yourself. Be sure this is acceptable to the insurance company.

Another alarm system which might appeal to you is the sort which is actually attached to valuable items. Anyone trying to remove these items without the key, trips off the alarm. They are a common feature in hi-fi shops and many saddleries now attach them to the D-rings of saddles.

## Guard Dogs

Although it may be tempting to leave a vicious dog loose in the yard or shop overnight, you mustn't, as you could find yourself having to pay damages to anyone who gets bitten (yes, even a thief!). Anyway, a dog loose in the shop would negate your alarm system. Don't let your big dog wander about the shop during the day, either. It will be off-putting to customers who do not know you and will prevent others combining 'walkies' with a buying trip.

Your best defence against pickers-up of unconsidered trifles when the shop is open is to arrange your displays in such a way that people cannot disappear from sight behind them – and to be sure that there is always an assistant in the shop in sight. (The companies who sell from stands at big shows often employ people called 'spotters' to wander around to watch for thieves.) If it is not possible to arrange displays to avoid dead-spots, some strategically placed mirrors will help you to keep an eye on people.

## Closed Circuit TV

Best of all is a closed circuit TV system. Its main value is as a deterrent, so the monitor sets and cameras must be immediately visible. Notices stating that the shop is protected in this way also help. You rent these systems on a weekly basis. For firms that deal with this, burglar alarms and all other aspects of security, look in the Yellow Pages under 'Security'. When choosing a security company to supply and fit your alarms, it is best to use a member of NACOSS (National Approved Council for Security Systems).

These firms can also help with the selection and installation of safes. Even if your insurance company does not insist, you should have one, for it is prudent to make a point of removing large denomination notes from the till at regular intervals. The best type is the 'hole in the floor' sort which has a little hole through which you poke the rolled notes, then you need only open it when you are ready to go to the bank. Put cheques and credit card receipts into it as well, for although they are less at risk than cash, they might be snatched from the till with a handful of notes.

## Safeguarding Cash

I mentioned earlier that, for the sake of your overdraft, you should bank your takings each day. You should do it for the sake of security also, because a determined thief will get into your safe one way or another. They might rip it out of the wall or floor, or they might decide it is easier to batter you into opening it for them.

The other reason you should visit the bank every day is that it is easier for all the parties concerned with your accounts if your bankings correspond with your till rolls. If anyone other than you uses the till, it is a good idea to empty it of all but its float at some point during the day, or more than once if you are taking a lot of money. Take a sub-total from the register and make sure the cash, cheques and credit card vouchers correspond. This is the best way of checking whether you have light-fingered helpers and is the best deterrent against this problem.

Go to the bank yourself. It is neither wise nor fair to get another member of staff to do it unless you are a small female and you send a large male. From the day you open your shop (and particularly in the early weeks, for this is when you are most likely to be observed for your 'easy job' potential), vary your route to the bank, vary the time you go and, if you can, vary your vehicle. Don't walk about with bank books and money obvious in your hand; put it in different bags, preferably shopping bags. Make sure that your insurance covers you for cash in transit to the bank as well as on your premises.

Arrange to use the bank's night-safe so you can get your money off your premises at weekends, bank holidays or late in the afternoon. You pay a small fee and the bank gives you a key to the night-safe and some leather pouches to put the money in. The pouches are also locked, and you are the only one with a key to them, so you go to the bank when it is open, ask for your pouches, open them and pay in as usual.

Pay your staff's wages by cheque. (You can make this one of your conditions of employment.) It saves you the chore of having to make up the right amount in cash and reduces the risk of keeping cash on the premises. It also simplifies your accounts and, with luck, leaves the money in your bank a little longer while the cheques are clearing!

## Cheques and Credit Cards

Unfortunately, this situation works both ways, for you also have to wait for clearance on your customers' cheques. But since most people don't carry large quantities of cash, you will lose a lot of business if you don't take them. Just refuse to accept them without a cheque card. These only guarantee cheques *up to* £50 (or the stated sum) and not, as some people think, the first £50 of larger cheques, so you must keep to this top limit. You must see the cheque being signed (a second time on the back if necessary) and must check the signature with the card. The code number must be written on the back of the cheque by you or your staff, *not the customer*, and it does no harm to add the card's expiry date as well.

The safest way to accept payment for goods is by credit card. They mean less cash in the till to tempt sticky fingers. There is absolutely no point in anyone stealing the completed vouchers, as they are imprinted with your shop's name and code number and are therefore of no value to anyone else. They cannot be endorsed to a third party, as stolen cheques can, and, provided you have kept to the rules laid down by the credit card company, you are guaranteed payment even if it turns out that the card was stolen or the card-holder was over their credit limit.

The rules are simple. With the hand-operated machines, you have what the card company calls a 'floor limit'. You must not process payments over this limit without telephoning for authorisation and you will be given an authorisation number to put on the slip. With the machines which are connected to the card company by telephone, these floor limits do not apply, and you can accept payments for any amount unless the telephone process refuses it.

Another advantage of cards over cheques is that you do not have to wait for clearance before you can be sure the money is in your bank. You pay the manual vouchers into your bank and the money is credited to your account immediately. The telephone transactions are automatically credited to your account the following day.

All this costs you is a small joining fee and a monthly service charge. This varies according to the volume of business you do, but it is normally a little over 2%. Incidentally, if you are selling by mail order and want to accept postal or telephone credit card business, you have to have an additional agreement as this is not covered by the normal agreement.

The security aspect is a major advantage with credit cards, but there is another just as important. Customers are more likely to buy expensive items such as saddles from you if they can pay by credit card, so it could give you the edge over the shop down the road which doesn't take them.

## Credit

Now that credit cards, debit cards and bank guarantee cards are so common, there is absolutely no reason why you should let *anyone* run a credit account. Quite apart from all the hassle associated with getting your money out of people and the inevitable bad debts when you fail with some of them, and the fact that they are having the use of *your* money until they pay, don't forget that you are liable to pay VAT when you invoice, not when you get paid.

The people most likely to ask for credit are the very ones

you should refuse – small riding schools or livery stables, both of which are notoriously bad payers. The only situation in which you might allow credit is where you sell feed on a large scale and deliver it to such establishments – and even in that situation you might want them to pay in advance. For other items from the tack side of your business, there are many potential problems. In any case, how can you police such an account? Who has authority to sign for goods? Will that authority be honoured at bill-paying time? The easiest way to deal with this whole area is to display one of those jokey cards which says 'Credit will only be given to persons over eighty with both parents present' or 'Please do not ask for credit as a smack in the mouth often offends'.

# 9
# *Advertising and Sponsoring*

Advertising is a legitimate business expense and is thus tax deductible, but unless you are advertising a specific item for sale, it is very difficult if not impossible to assess its impact and therefore its value.

## *Advertising in the Press*

Your first thought will probably be that you should advertise in *Horse and Hound* or the monthly magazines. But before you do so, take the trouble to examine these publications and see how many retail outlets advertise in them, and exactly what it is they are advertising. On the whole, you will find they are either specialists, or offering a mail order service.

The exceptions are the Show editions. This is because extra copies are printed and the people who buy these editions tend to keep them all year instead of throwing them away when they've read them. This means that for the cost of one insertion (or three in *Horse and Hound*) you effectively get a full year's coverage.

Display advertisements in such magazines are very expensive. If their sole purpose is to tell your *local* customers you exist, they are a waste of money. If you've been around horses long enough to see the potential of a tack shop, you should be fully aware of the speed and coverage of the equine grape-vine and you should know that 90% of your potential customers over a twenty-mile radius will know of your existence, stock and price range within a fortnight of your opening.

## PRESS RELEASES

By all means make a splash to announce your opening and any special events such as Sales, but you will probably get best value for money from your local newspaper. If you handle it right, you should be able to get some free editorial coverage.

Local newspapers are interested in local events and local people, so you have a double advantage if you have lived in the district for any length of time. You can even tell all the readers what a splendid person you are under the guise of a potted life history. Rather like preparing a CV to send to prospective employers, you play up the good bits and keep quiet about the rest: 'Mr Fred Bloggs has lived in Blanktown for the last six months after spending most of his life in Grungeshire where he is well known in the horse fraternity for his enthusiasm in supporting local show-jumping events. (You don't mention that you've never had a clear round, let alone won anything!) He is now applying his expertise for the benefit of the riders of Blanktown by opening a tack shop in Station Road', and so on.

There are two ways to do this. Either you put on your PR hat and send out a proper press release (headed 'Press Release') with a story ready written and a couple of captioned photographs of you doing something exciting on a horse, or standing by a horse which is covered in rosettes (it doesn't even have to be your horse as long as the caption is noncommittal – 'Mr Bloggs at last year's Grunge County Show') and a professional photograph or two of the new shop. The average editor will be delighted with this ready-made copy, especially if you have photographs handy to go with it. The alternative is to telephone the paper, tell them when you are going to open and ask if they'd like to send a reporter and a photographer.

The paper will probably suggest that you place an advertisement on the same page as their editorial. Don't – people will assume the whole thing is an advertisement placed by you, and this reduces its value.

Keep sending the press releases, with captioned photographs, every time you do something that constitutes good copy. You ought to get a mention a couple of times a year this way, especially if you remember that a lot of things that seem old hat to you could be an interesting novelty to the editor: 'A team from the Blanktown Riding Club has entered for the Golden Horseshoe Ride . . . This is one of the fastest-growing branches of competitive riding. Mr Bloggs of Bloggs' Tack Shop who are now stocking a full range of long distance riding equipment said . . .'

Incidentally, do send copies of your press releases to your local radio station. They are less likely to use them than the newspaper, but you might be lucky.

Although you shouldn't advertise on the same page as your editorial mentions, there is no reason why you shouldn't advertise in the local paper on a regular basis if the rates are right. The cheapest way is to do it for a specified period of several weeks. The acknowledged best places are the front or back pages and, although these are obviously in high demand, you might be able to arrange it for a long period. Talk to the advertising department and see how amenable they are, not forgetting that most of them get paid commission for the amount of space they sell.

## Other Ways to Advertise

There are other, cheaper ways to advertise. One is to pay for advertisements in local show programmes. See if your local Riding Club will take an advertisement for their newsletter. Better yet, see if they will let you pay for the postage on a newsletter in return for including an insert. Offer them a special deal on something like safety tabards – with your name on them.

Put posters up in riding schools' waiting rooms. Don't forget the Yellow Pages. Give away bumper stickers, or give away a free T-shirt to the first fifty customers – 'I was first at Bloggs' Tack Shop'. Give every customer a printed carrier bag and make sure it's a stout one that will last a long time.

Don't ask local horse professionals like vets, farriers or free-lance teachers to spread the word. All are independently minded people who could take offence and tell people what a nerve you've got, which is not what you want.

Do get a good sign-writing job done on your van. Then keep it clean and taxed – you don't need newspaper coverage of your fine! And don't allow it to be driven around the district in a manner that alienates people or exposes you to bad editorials.

Take every opportunity to give talks to Riding Clubs or evening classes on horsemastership. The choice of topics is endless – 'Draw-reins as a Schooling Aid'; 'Road Safety'; 'Recent Developments in Saddle Manufacture'. (You do know your subject, don't you?)

Beware of marketing men who want to sell you gimmicky ideas like balloon races. Advertising is only effective if it reaches its target audience. Balloon finders in Spain are not a lot of use to you.

Advertising attached to non-equine charity events is a mixed blessing. Some of these charities do hold horse shows, but there is a school of thought that says you should be prepared to pay extra because it is for charity. If you really want to make a donation, this is a tax-deductible way to do it. If you don't, check out the prices first.

You might also think of acquiring an Internet web site. This should be in addition to, rather than an alternative to, advertising in magazines, but it does give you an easy way of updating details of your goods. Because many businesses also sell direct from their web sites, I have dealt with it in detail in the next chapter.

## Sponsorship

This is considered to be a form of advertising and so goes under that heading in your accounts.

All will be well as long as the taxman does not suspect that you are using it as an excuse to subsidise your own hobby animals. If you are using one of your family beasts you will have to be careful. Make sure that it has a name that includes your trading name, in such a way that show announcers will be sure to use it (Bloggs' Tack Betty?). Make sure it gets to enough shows to carry the name far and wide. And make sure it is good enough to win regularly – 'Bloggs' Tack Betty has fallen at the first fence again' isn't going to do your image a lot of good.

All of this also applies if the animal does not live *en famille* but is kept by a professional. And do ask yourself at intervals if it is a worthwhile exercise, or an expensive indulgence that is eating up all your profits.

Another expensive indulgence is that of sponsoring classes at your local show by providing prizes or prize money. It is traditional to ask local tack shops to do this and it is easy to be persuaded into doing more than one class. Before you know where you are, you're committed to £200, and all you get for it is a couple of mentions in the programme when you could have had a full-page advert in the programme for £40.

The best idea I've encountered on this theme is to give vouchers, redeemable for goods in your shop. Twenty £10 vouchers, one for the winner of each class, does not even cost you £200, it costs the wholesale price of £200 worth of goods. And then only if each winner (a) bothers to come and use it and (b) takes an item selling for exactly £10. Most wait until they need something (during which time you've had the use of the money) and will spend much more than £10. If they have come from outside your usual catchment area, even better.

Another way to get good value for money is to sponsor a County or Area event. Something like 'The Bloggs' Tack Grunge County Handy Pony Championships' or something a little rarer like a driving championship. £350 will give five qualifying classes at £50 and a final at £100 – and it will give you a very grateful organising body who will bend over backwards to sing your praises to their members. They'll send reports to the local papers and the national horse press, give you very cheap, if not free advertisements in the programmes and cheap, if not free, stand space at their shows. Which you utilise by taking along items likely to appeal to the competitors, who already think you're marvellous!

Be sure your attitude does not disabuse them of this notion.

The one essential for any retailer is to be helpful, and the more competitive the trade, the more helpful you need to be. Another tack shop here, now defunct with no surprise to any of us who ever used it, was owned by a terse and rather ignorant know-it-all. In addition to his generally off-putting attitude, he tried to cash in on the growing craze for side-saddle and was rash enough to try to tell me what was what, when he knew full well who I was, and should have realised that I knew more about it than he did.

So another rule is, know your customers, know their specialities and expertise and don't alienate them by talking down to them. A little research can work wonders and deferring to experts isn't creeping, it is good business sense.

# 10

# Sales, Specialities, Mail Order and Show Stands

## Sales

Without a crystal ball you have no way of telling how many of an article you can sell, nor even if you can sell it at all. Wholesalers' representatives will tell you what goes well in other shops, but that does not necessarily mean it will for you. Nor have you any way of telling from year to year how many seasonal items like rugs will sell. So you will find that 'dead' stock will accumulate. It will get less as time goes by and you build up experience of the buying habits of your local customers, but even after many years, there will be items which will confound you by refusing to budge from the shelves. Weather conditions too, will play their part in how well seasonal items sell.

Anyway, there you will be, in the spring and autumn, looking at unsaleable items and wondering what to do with them. You have the choice of either putting them away until it is their turn again, and hope the new season's prices have gone up enough to compensate you for having your money tied up, or selling them cheaply to get some working capital back to work.

At which point it begins to be a good idea to hold a spring/autumn Sale to get rid of these items and get your capital back to work. There are plenty of horse owners who will see their advantage in buying a rug or whatever for next winter at a reduced price, especially if you point out what next season's prices are likely to be. All you have to do is

75

achieve the right balance between making a bit of profit and making the item attractive to the buyers. I would suggest that a 20% reduction from the normal price will attract the customers, cover your overdraft costs and allow a bit for overheads.

Announce the Sale as being for two weeks with an extra Saturday thrown in, and if you haven't got rid of all your Sale items by then, try one last Saturday at an additional reduction or put a 'Make me an offer' label on them, having first decided what is the least you will accept.

Incidentally, the law now says that prices displayed as 'original' prices (you know – 'Was £60, now only £40') must be the genuine price of the item for the last six months. No marking it up for a couple of weeks before the Sale to make your Sale price look extra generous!

Wholesalers and manufacturers also suffer from the problem of unmoving stock, and they also have periodical clearouts and offer retailers such items at reduced prices. This is not the same thing as many retail outlets do at Sale time of buying in special goods to boost the display of generosity. You could do that, but it is not a good idea if those special goods are of inferior quality. If you have been painstakingly building a reputation for quality goods, you will damage that reputation by selling 'tat' at Sale time – and what are you going to do with the unsold 'tat' when the Sale is over?

## Specialising

One way of ensuring that your selling power has not reached saturation point in the district is to widen your district by specialisation. Choose an area, preferably a competitive one, which is just obscure enough to prevent the average shop carrying a good range of items for its devotees, and open a department that stocks everything those devotees could want. Find out just what they do want by consulting the acknowledged expert and announce that fact in your advertisements.

It would be rash to concentrate on that speciality to the

exclusion of your run-of-the-mill customers, for you may find that your chosen speciality proves to be a passing fad. Whatever happened to Western Riding for instance? At one time it was all the rage and there were demonstration teams and classes at every show. You hardly ever see it now, so the lofts of the horse fraternity must contain many collections of unsellable Western saddles and bridles as well as clothing.

The knack is to detect an enthusiasm for something new before it gets so popular that too many other shops jump on the bandwagon, and set yourself up as THE specialist supplier. Help it along a little with those Area Championships I mentioned in the previous chapter; advertise in the horse magazines, write letters to them or articles for them, to keep the momentum going – but keep a weather eye open for waning interest so you don't get stuck with a lot of unmovable goods.

The alternative is to specialise in one particular aspect of an ongoing activity. Top hats, for instance, are worn in the hunting field, in the dressage arena, by competitors in the night classes at the International; by side-saddle and driving enthusiasts. There aren't enough of these people to make top hats a standard item in every tack shop, but they have to get them from somewhere, so it has to be from London – or a specialist.

## Mail Order Selling

Whatever your speciality, you can't expect all your customers to come to you, so you must either take your wares to them by travelling the show circuit (of which more later) or you must sell by mail order. This is not an area that is confined to specialist items. There are many horse owners who live in remote or sparsely horse-populated areas who have no tack shop within reasonable distance, so they must buy most of their supplies through the post.

There are three ways you can approach mail order selling. The first is known as 'off the page' selling, where you place an advertisement offering specific items for sale, and the customers write or telephone to order those only. The second method is where you produce a catalogue of your wares, which you then post to potential clients. You either find them by advertising or by purchasing a list of known buyers of equestrian products. Organisations who supply such lists are known as list brokers, and the only one I have been able to find who can produce a list of *horsy* people is Equestrian Management Consultants Ltd., a side-shoot of BETA. It is easier to buy lists of the less specific category; 'people interested in country pursuits'.

In fact you don't normally buy a list as such, but a set of labels to stick on your envelopes. These are provided under the condition that you don't copy the addresses for further use, and it is normal practice for list brokers to include some 'jokers' in the list, so they will very quickly find out if you do try this. But it really isn't worth doing – think of the cost of paying someone to copy all those addresses, when buying the list in the first place will only cost about £100 per 1000 addresses. The other side of this coin, of course, is that if you do build up your own mailing list of people who buy equestrian items, you can sell your list to a list broker. The more specific your list, the easier and more lucrative it is to sell.

Selling through catalogues is not a cheap option. The more elaborate the catalogue, the more it costs to produce. Don't even *think* of colour photographs unless you are sure you are going to sell hundreds of items – no printer will print such a brochure in small quantities and your initial outlay will be enormous. (Somewhere in the region of 50p to £1 each with black and white photos, much more with colour.) You will also have to pay postage, and employ someone to 'stuff' the envelopes, add labels and so on. The normal method of doing this is to use a 'mailing house', who will tackle the whole job, including supplying envelopes. The catalogues go straight to the mailing house from the printer, and they pass on any spares to you after the main mailing has been done.

Traditionally the best place for advertising mail order items is the national horse magazines, Which one(s) you choose will depend upon whom it is you want to reach. For hunting and racing people it has to be *Horse and Hound*; for items likely to appeal to children, it has to be *Pony*, for general riders there are *Horse and Rider* and *Riding*. For the more specialised interests, it could also be the smaller circulation magazines aimed at that particular audience, but you should use one of the general magazines as well.

Whether you list all your wares or offer a catalogue, no magazine will accept your advertisements unless you comply with their 'Mail Order Protection' (MOPS) requirements. This is a scheme organised through the Periodical Publishers Association, who have a booklet called *Guidelines for new advertisers*. (see Appendix 6 for their address). The scheme basically concerns quality of goods and requires you to operate a 'return goods, get your money back' guarantee. (If you advertise in national newspapers, the scheme is slightly different.)

## Websites and the Internet

The third way of selling by mail order is through the Internet. To do this you need a website, which you acquire and run either by doing it yourself (not a good idea unless you are extremely computer literate and prepared to buy an expensive computer which then has to be left running and connected to an open telephone line all the time) or you go to a website 'provider' and 'host' who will do it all for you. This reduces the hassle factor and doesn't have to be desperately expensive*.

Your website can either be a simple one which says how wonderful you are, that you have a catalogue and will send it if people send you an email message, or it can contain your catalogue, which will of course mean more pages. You will obviously have to update it regularly. You can then opt to give a phone number so customers ring you to place their order, or You can do the whole thing electronically.

*At the time of writing, I was told it should cost about £50 per page, plus less than £100 per year for the 'space' on a host computer.

This latter option is not as simple as it sounds, and not to be taken on lightly, as there are all sorts of horrible complications. The least of these is that you need to use a host with a secure server to process credit card orders. Then your site has to be much more complex, with a built-in ordering facility. Finally, because the worldwide web is just that - world wide, you cannot restrict yourself to the UK only. Willy nilly, you will find orders have come in from all over the world, and that opens another big can of worms - export and import regulations, customs declarations (and the problems of various countries which may decide to place embargoes on certain products or materials), packing and postage, dealing with returns, currency rates and all that goes with them.

I have listed a few companies who will help with setting up websites in Appendix 6.

## Delivery Times

One joy with mail order is that you can say 'Allow twenty-eight days for delivery', which could mean you don't even have to hold a lot of stock. The MOPS rules say you must hold 'sufficient to meet the likely demand', but given that most wholesalers now guarantee forty-eight hours delivery, those stocks do not need to be enormous. In other words, you do not need to order the goods from your supplier until you have sold them - and have been paid by your customers!

Even without the twenty-eight day rule, you do not send the goods until the customer's cheque has cleared. With credit cards, however, this problem does not arise. I was told by one retailer that their mail order sales increased dramatically when they started taking credit card telephone orders. (Don't forget that you need an additional agreement with the credit card company to do this.) Put yourself in the customer's place - no forms to fill in, no letters to write and post, just pick up the telephone. You can even talk to someone about the item you want, to ask if it is suitable.

Whichever way you reach your customers, you will have to

budget for continuous advertising, packaging materials, goods in transit insurance, postage or carriage charges, and the inevitable returns.

## *Show Stands*

You may have perked up when I mentioned these earlier as a good outlet for specialities, and thought it would be nice to go round the big shows. Sorry – not such a good idea. The regular exhibitors make a success of it because they work damn hard at it and are very professional in the way they go about it. Even so, many of the familiar names are gradually dropping out.

For a start it is an expensive operation to mount. You need a properly designed show unit, preferably of rigid construction (or you need to be good at erecting tents), with the displays carefully thought out. You need a vehicle to transport it; you need a caravan for the staff to live in, and something to tow that as well as another van to ply between shows and base with stock replenishments. You will also need an amenable insurance company because pilferage at shows is rife.

You will have to book your sites months or even years ahead (someone mentioned a seven-year waiting list for one of the major events) and pay a great deal for those sites. You won't get much change out of £1,000 at most County shows and the top five events are in excess of £1,000. Not to mention that if you want to take stand space at one of the big exhibition centres such as Olympia or Earls Court, you have to be prepared to go along with the requirements of the trade unions which control stand assembly.

You need the right staff, with iron constitutions and stable domestic lives, for if you are doing the whole circuit, they will leave home in April and will hardly get back until October. They have to work in all weathers, sunshine or rain, dry ground or quagmire, while living out of a caravan. Apart from setting up and dismantling the stand at each show, with the attendant lugging of stuff in and out of lorries, they have to spend a lot of time in wet weather cleaning their stock.

As a beginner, or as a small general tack shop, you would be well advised to stick to small local shows. Your stand need be no more than a tent with trestle tables and a couple of revolving clothes racks. Your site rent will be well under £50 unless you insist on exclusivity. Riding Club shows may even be free. By all means take saddles, though you are unlikely to sell one unless a customer has specially asked you to bring them. A show, after all, is a likely place to try on saddles if it is not too noisy and horse-exciting.

The likeliest items to sell are the small ones, or things that break at shows, like reins, lead ropes or headcollars. Have a look at the schedule before you go, to see what the classes are, and pack accordingly. A heavy concentration of children's classes and you take pony tack, children's jackets and hats and 'Thelwell' giftware. Jumping classes means you take nose-bands and martingales, rubber reins and over-reach boots – and whips.

Check the weather forecast too and rejoice if it is going to rain later in the day. Then you take rubber boots, rainwear and umbrellas and sell to all the people who didn't see (or listen to) the weather forecast, or forgot their macs.

But don't expect to make your fortune at it. You probably won't sell that much, or only small, low-profit items. The best attitude is to consider any profits as a bonus, regard the whole thing as an advertising exercise, then relax and enjoy the show.

## Refreshments for Customers

If you have the space, you might think of adding a coffee shop, especially if your customers come a long way to reach you. Not only will they be grateful for a decent cup of tea or coffee (drinks machines tend to be more hassle than they are worth and we all know what the drinks taste like), being able to sit down and relax will put them in a better frame of mind for spending money. Given a pleasant location, or a lot of passing trade, you may even find non-horsy locals will come in.

As with any additional and different type of item, you will need to obtain planning permission for a coffee shop, and there are other complications such as hygiene regulations. In view of the latter, and the fact that the restaurant business is notorious for losing inexperienced peoples' money for them, the best way to provide this facility for your customers and earn some extra money is not to run it yourself, but to rent the space to a professional.

# 11

# *Adding a Tack Shop to Your Stables*

If you are running a stables that has reached its maximum profit level, or if you have a riding school where your pupils have to go twenty miles to buy their riding clothes, it might seem a good idea to add a shop. So it is, but you should not rush into it without careful thought, planning and research.

## *Planning Permission*

You must have 'change of use' permission to create a shop in another type of premises, and since you will almost certainly be in a rural location, you may well be refused on the grounds that opening a shop constitutes 'bringing the town into the country'. If this comes up, you could plead for a two-year licence and hope the neighbours do not complain about the increased volume of traffic when you have to renew. You will need adequate car parking facilities, for cars left in the road will not please the neighbours either.

The next point is whether your lease or tenancy agreement will allow it, or whether you can negotiate a variation.

## *The Competition*

Next, and most important, how close is the competition and how narked will they be? The least they will do is object to your planning applications; they are also likely to put pressure on the wholesalers and manufacturers to restrict your supplies. Not that they will have to do the latter – the whole of the saddlery trade objects to what they think of as 'cowboy' retailers, and you could find it very difficult to get stock at all. Even if all you are thinking of doing is stocking clothing (which they also stock), you might find they will refuse to sell you anything else and refuse to do your repairs.

## Basic Considerations

Where are you going to put the shop? In what sort of building? Pupil customers will be happiest if it is attached to the indoor school or waiting room, or accessible from either without going out of doors. Your insurance company will be happiest if it is a solid brick building, partly because of the fire risk and partly because tack thieves like big stables and they'll love one with a shop full of new goodies even more.

Who will be your customers? If you intend to serve the entire district, see Chapter 2. If you intend to serve no more than your school pupils, you will not need to stock anything other than clothing, boots, hats, gloves, whips and books. The level of sophistication of all these items will be dependant upon the age of your customers and the level of teaching you offer. If you run shows regularly, add the items that need emergency replacements as listed in Chapter 10.

Who is going to serve customers? You cannot do it on a basis of letting whoever is handy have the key to open up when a customer appears. Nor can you leave it unlocked and unattended if you don't want things to 'walk'.

You must have definite 'open' hours, when the shop is staffed. In a school situation these could be a couple of hours in the early evening and all day Saturday and Sunday, in which case one part-timer could probably cope. But for the sake of security, there must be one person with ultimate responsibility for the takings and stock, and that person should hold the key.

The 'No Credit' rule will have to be rigidly enforced.

Whether it's 'Mummy will pay for it when she collects me', or 'I haven't got my purse, can I pay you next time?', the answer must be 'I'm sorry, the boss won't let me'. That applies equally to other staff, working pupils and freelance instructors as well as the public.

The other rule that must not be broken is that stock must not be taken from the shop for use in the stables without being listed in a special book and signed for. Then, at the end of each month, the shop should issue an invoice to the stables, and the stables must pay it. Your accountant will emphasise that the shop should be run as a separate entity and it should have a separate bank account. That way it won't complicate the other VAT problem, nor the other staff situation (see Chapter 7).

Best of all, it will be easy to see if it is just another drain on your resources, or a profitable adjunct to your business.

# Appendices

# Appendix 1 - *Cash Book*

*(Income)*

| Date | Total | Zero-rated | Std-rated | Repairs | VAT |
|---|---|---|---|---|---|
| Jan 2 | 592.13 | 146.48 | 256.28 | 123.00 | 66.37 |
| Jan 3 | 691.39 | 172.12 | 414.64 | 27.30 | 77.33 |
| Jan 4 | 984.35 | 233.76 | 638.80 | 0.00 | 111.79 |
| Jan 5 | 681.15 | 78.52 | 565.84 | 18.80 | 17.99 |
| Jan 6 | 1,102.79 | 334.76 | 596.45 | 57.20 | 114.38 |
| Jan 7 | 4,010.14 | 958.50 | 2,409.14 | 188.00 | 454.50 |

*(Expenditure)*

| Date | Payee | Cheque No. | Amount | Stock Zero-rated | Stock Std-rated | Wages | Motor | Other | | VAT |
|---|---|---|---|---|---|---|---|---|---|---|
| Jan 2 | Shop-Shield Ltd | Stdg Order | 305.50 | | | | | Alarm Hire | 260.00 | 45.50 |
| | C. Perkins | Stdg Order | 3,750.00 | | | | | Rent | 3,750.00 | |
| | Blanktown Council | Stdg Order | 350.00 | | | | | Rates | 350.00 | |
| | Loan repayment | Stdg Order | 2,600.00 | | | | | Loan | 2,600.00 | |
| | Bank Charges | Stdg Order | 1,206.72 | | | | | Charges | 1,206.72 | |
| Jan 6 | Central Elec Board | 284 | 492.17 | | | | | Electricity | 468.73 | 23.44 |
| | British Telecom | 285 | 433.76 | | | | | Phone | 369.16 | 64.60 |
| | Post Office | 286 | 27.80 | | | | | Parcels | 27.80 | |
| | Green | 287 | 250.00 | | | 250.00 | | | | |
| | Black | 288 | 250.00 | | | 250.00 | | | | |
| | Blue | 289 | 280.00 | | | 280.00 | | | | |
| | Petty cash | 290 | 50.00 | | | | | Petty cash | 50.00 | |
| | SEIB | 291 | 1,783.96 | | | | | Insurance | 1,732.00 | 51.96 |
| | BH&B | 292 | 219.18 | | 186.54 | | | | | 32.64 |
| | Shires | 293 | 1,048.27 | | 736.87 | | | | | 128.95 |
| | Jeffries | 294 | 445.46 | 182.45 | 379.12 | | | | | 66.34 |
| | Bloggs Garage | 295 | 424.26 | | | | 390.86 | Service | | 33.40 |

# Appendix 2 - *Petty Cash Book*

| Income | | | Expenditure | | |
|--------|--------|-------|--------|--------|-------|
| Jan 6 | Chq No 29 | 50.00 | Jan 1 | Bal b/fwd | 2.24 |
| | | | Jan 2 | Coffee, sugar | 6.56 |
| | | | Jan 6 | Stamps | 5.20 |
| | | | | Envelopes, pens | 1.77 |
| | | | | Dustmen | 5.00 |
| | | | Jan 7 | Elastoplast | 1.92 |
| | | | | Milk | 1.90 |
| | | | Jan 9 | Sugar, biscuits | 3.95 |
| | | | Jan 13 | Milk | 1.90 |
| | | | | | 30.44 |
| | | | | Bal c/fwd | 19.56 |
| | | 50.00 | | | 50.00 |

# Appendix 3 - *Cash Flow Forecast*

|  | Jan | | Feb | | Mar | | Apr | |
|---|---|---|---|---|---|---|---|---|
|  | Actual | Budget | Actual | Budget | Actual | Budget | Actual | Budget |
| **Income** | | | | | | | | |
| Zero-rated | 7773 | 8000 | 10432 | 8000 | 13040 | 10000 | | 8000 |
| Standard-rated | 19524 | 20000 | 14496 | 20000 | 18122 | 25000 | | 20000 |
| Repairs | 1656 | 2000 | 1540 | 2000 | 2074 | 2500 | | 2000 |
| VAT | 3707 | 4000 | 2808 | 4000 | 3534 | 5000 | | 4000 |
| **Total Income** | 32660 | 34000 | 29276 | 34000 | 36770 | 42500 | | 34000 |
| **Expenditure** | | | | | | | | |
| Stock | 13018 | 15000 | 11587 | 15000 | 17249 | 15000 | | 15000 |
| Staff wages | 3120 | 3120 | 3120 | 3120 | 3120 | 3120 | | 3500 |
| Rent | 3750 | 3750 | 3750 | 3750 | 3750 | 3750 | | 3750 |
| Rates | 350 | 350 | 350 | 350 | 350 | 350 | | 400 |
| Electricity | 1492 | 1500 | | | | | | 450 |
| Telephone | 434 | 500 | | | | | | 500 |
| Insurances | 1784 | 1800 | | | | | | |
| Motor expenses | 424 | 400 | 185 | 200 | 212 | 200 | | 400 |
| Repairs & maint | | 250 | | | 136 | | | 250 |
| Accountancy | | | | | | | | |
| Security | 306 | 300 | | 300 | 300 | 300 | | 300 |
| Contingencies | | 250 | 627 | 250 | | 250 | | 250 |
| General & p. cash | 128 | 250 | 176 | 250 | 289 | 250 | | 250 |
| Loan repayments | 2600 | 2600 | | | | | | 2600 |
| Bank Charges | 1207 | 1250 | | | | 1250 | | |
| VAT | 1954 | 2000 | 1911 | 2000 | 2372 | 2000 | | 2000 |
| Proprietors' drawings | 4000 | 4000 | 4000 | 4000 | 5000 | 5000 | | 5000 |
| **Total Expenditure** | 34567 | 37320 | 25706 | 29220 | 32778 | 31470 | | 34650 |
| Month's Balance | (1,907) | (3,320) | 3,570 | 4,780 | 3,992 | 11,030 | | (650) |
| Cumulative Balance | (1,907) | (3,320) | 1,663 | 1,460 | 5,655 | 12,490 | | 11840 |

Notes
1. VAT has been calculated at a standard rate of 17.5%.
2. In the balance lines, the accounting convention of showing minus figures in brackets has been used.
3. It has been assumed that March, June, September and December have 5 weeks and that all other months have 4 weeks.
4. Contingencies are unexpected items which cannot be specified in advance.
5. General items are those small items which are too small to specify, such as postage, petty cash etc.
6. Proprietors' drawings are the money a self-employed person takes as 'wages'.

| May (Actual) | May (Budget) | Jun (Actual) | Jun (Budget) | Jul/Aug/Sep (Budget) | Oct/Nov/Dec (Budget) | Jan/Feb/Mar (Budget) | Apr/May/Jun (Budget) | Jul/Aug/Sep (Budget) | Oct/Nov/Dec (Budget) |
|---|---|---|---|---|---|---|---|---|---|
| | 8000 | | 10000 | 36000 | 36000 | 40000 | 40000 | 40000 | 40000 |
| | 20000 | | 25000 | 65000 | 65000 | 72000 | 72000 | 72000 | 72000 |
| | 2000 | | 2500 | 6500 | 6500 | 7200 | 7200 | 7200 | 7200 |
| | 4000 | | 5000 | 17000 | 17000 | 14000 | 14000 | 14000 | 14000 |
| | 34000 | | 42500 | 124500 | 124500 | 133200 | 133200 | 133200 | 133200 |
| | 15000 | | 15000 | 50000 | 50000 | 55000 | 55000 | 60000 | 60000 |
| | 3500 | | 3500 | 10500 | 10500 | 10500 | 12000 | 12000 | 12000 |
| | 3750 | | 3750 | 11250 | 11250 | 11250 | 11250 | 11250 | 11250 |
| | 400 | | 400 | 1200 | 1200 | 1200 | 1350 | 1350 | 1350 |
| | | | | 1500 | 1750 | 1750 | 1500 | 1600 | 1750 |
| | 500 | | 500 | | | | | | |
| | | | | 1800 | | 2000 | | 2000 | |
| | 200 | | 200 | 1000 | 1000 | 1000 | 1000 | 1000 | 1000 |
| | | | | 250 | 250 | 250 | 250 | 250 | 250 |
| | | | 7500 | | | | 7500 | | |
| | 300 | | 300 | 1000 | 1000 | 1000 | 1000 | 1000 | 1000 |
| | 250 | | 250 | 750 | 750 | 750 | 750 | 750 | 750 |
| | 250 | | 250 | 750 | 750 | 750 | 750 | 750 | 750 |
| | | | | 2600 | 2600 | 2600 | 2600 | 2600 | 2600 |
| | | | 1250 | 1250 | 1250 | 1250 | 1250 | 1250 | 1250 |
| | 2000 | | 2000 | 6500 | 6500 | 7500 | 7500 | 8000 | 8000 |
| | 5000 | | 6250 | 16500 | 16500 | 18000 | 18000 | 20000 | 20000 |
| | 31150 | | 41150 | 106850 | 105300 | 114800 | 121700 | 123800 | 121950 |
| | 2850 | | 1350 | 17650 | 19200 | 18400 | 11500 | 9400 | 11250 |
| | 14690 | | 16040 | 33690 | 52890 | 71290 | 82790 | 92190 | 103440 |

# Appendix 4 - *Stock Record Card*

Item .................................................... RRP ...............................................................

Stock No .................................................... Delivery ............................................................

Details .................................................... VAT ................................................................

Reorder level ...................................................

| Ordered Date | No. | Delivery Date | Invoice No. | Item Cost | Remarks | Stock Date | Check Qty |
|---|---|---|---|---|---|---|---|
| | | | | | | | |

Supplier ...........................................

Telephone ...................................

# Appendix 5 - *Deductible Expenses*

Any sums expended wholly and exclusively for the purposes of the business may usually be deducted in the computation of profits. These sums must be of a 'day-to-day' nature, and not for capital items such as buildings, shop fittings or vehicles. The list is intended to be a guide, not exhaustive.

a) Advertising and sponsorship
b) Loan interest
c) Insurance premiums
d) Redundancy payments
e) Wages, pensions and employer's Social Security contributions
f) Bad debts
g) Legal expenses
h) Rent and Rates
i) Repair and maintenance charges
j) Hire or lease charges on such items as vehicles, machinery or security equipment
k) Subscriptions to societies and periodicals
l) Other business expenditure – telephone, light and heating, postage and stationery, staff refreshments, furniture for staff rooms and offices, specialist working-clothes, feed and veterinary expenses for guard dogs or rodent-catching cats.

# Appendix 6 - *Useful Addresses*

This list includes all those mentioned in the text, and a few more useful contacts. For a full listing of wholesalers, you need the *British Trade Suppliers Directory* from the British Equestrian Trade Association (BETA) at Wothersome Grange, Bramham, Wetherby, W. Yorks LS23 6LY. Tel. 0113 289 2267 Internet: www.u-net.com (Also at this address: Equestrian Trade News and Equestrian Management Consultants Ltd.)

J A Allen, 45–47 Clerkenwell Green, London EC1R 0HT.
Tel. 0171 251 2661
(Publishers of equestrian books)

Allweb, The Bank, Nantmawr, Oswestry, Shrops SY10 9HN.
Tel. 01691 828729
Internet: www.allweb.co.uk
(Website creators/host)

Croner Publications Ltd., Croner House, London Road, Kingston-upon-Thames, Surrey KT2 6BR.
Tel. 0208 247 1175
Internet www.croner.co.uk
(Publishers of *Croner's Reference Book for the Self-employed and Smaller Business*)

*Equestrian Business News*, Bruce Publishing, PO Box 82, Wantage, Oxon OX12 7YU.
Tel. 01235 771770
Internet: www.brucepub.com

*Farm & Country Retailer* magazine, John C Alborough Ltd., Battisford Road, Ringshall, Suffolk IP14 2JA.
Tel. 01473 658006
Email:jca_pharm@compuserve.com

Guildsoft Ltd., The Software Centre, East Way, Lee Mill Industrial Estate, Ivybridge, Devon PL21 9GE.
Tel. 01752 895100
Internet: www.guildsoft.co.uk
(Distributors of MYOB accounting and other software)

Hartland Web Pages, 38 Coxheath Road, Church Crookham, Fleet, Hants GU13 0QG.
Tel. 01252 408877
Internet www.hartlana.co.uk
(Website creators/host)

Periodical Publishers Association, Queens House, 28 Kingsway, London WC2B 6JR.
Tel. 0207 404 4166
(Run mail order protection scheme for magazine publishers)

REMUS, PO Box 39, Hailsham, East Sussex BN27 2QL.
Tel. 0500 823290
Internet: www.remus.com
(Website creators/host)

Sage Software Ltd., Sage House, Benton Park Road, Newcastle-upon-Tyne NE7 7LZ.
Tel. 0191 255 3000
Internet: www.sage.com
(Producers of accounting software)

Shires Equestrian Products, M J Ainge & Co. Ltd., 15 Southern Avenue, Leominster, Herefordshire HR6 0QF.
Tel. 01568 613599
Internet www.shires-equestrian.co.uk
(Manufacturer and wholesaler of various items. They will also assist with catalogue production and setting up websites)

South Essex Insurance Brokers, South Essex House, North Road, South Ockendon, Essex RM15 5BE.
Tel. 01708 850000
(Give discounts to BETA members, and are a BETA recommended broker)

# Index